# A Practical Approach to
# Medical Image Processing

# Series in Medical Physics and Biomedical Engineering

Series Editors: John G Webster, E Russell Ritenour, Slavik Tabakov,
and Kwan-Hoong Ng

*Other recent books in the series:*

**Biomolecular Action of Ionizing Radiation**
Shirley Lehnert

**An Introduction to Rehabilitation Engineering**
R A Cooper, H Ohnabe, and D A Hobson

**The Physics of Modern Brachytherapy for Oncology**
D Baltas, N Zamboglou, and L Sakelliou

**Electrical Impedance Tomography**
D Holder (Ed)

**Contemporary IMRT**
S Webb

**The Physical Measurement of Bone**
C M Langton and C F Njeh (Eds)

**Therapeutic Applications of Monte Carlo Calculations
in Nuclear Medicine**
H Zaidi and G Sgouros (Eds)

**Minimally Invasive Medical Technology**
J G Webster (Ed)

**Intensity-Modulated Radiation Therapy**
S Webb

**Physics for Diagnostic Radiology**
P Dendy and B Heaton

**Achieving Quality in Brachytherapy**
B R Thomadsen

**Medical Physics and Biomedical Engineering**
B H Brown, R H Smallwood, D C Barber, P V Lawford, and D R Hose

**Monte Carlo Calculations in Nuclear Medicine**
M Ljungberg, S-E Strand, and M A King (Eds)

**Introductory Medical Statistics 3rd Edition**
R F Mould

**Ultrasound in Medicine**
F A Duck, A C Barber, and H C Starritt (Eds)

Series in Medical Physics and Biomedical Engineering

# A Practical Approach to Medical Image Processing

**Elizabeth Berry**
*Elizabeth Berry Ltd*
*Leeds, UK*

CRC Press
Taylor & Francis Group
Boca Raton London New York

CRC Press is an imprint of the
Taylor & Francis Group, an **informa** business
A TAYLOR & FRANCIS BOOK

CRC Press
Taylor & Francis Group
6000 Broken Sound Parkway NW, Suite 300
Boca Raton, FL 33487-2742

© 2007 by Taylor & Francis Group, LLC
CRC Press is an imprint of Taylor & Francis Group, an Informa business

First issued in paperback 2019

No claim to original U.S. Government works

ISBN-13: 978-0-367-45284-1 (pbk)
ISBN-13: 978-1-58488-824-6 (hbk)

### Library of Congress Cataloging-in-Publication Data

Berry, Elizabeth, 1961-
 A practical approach to medical image processing / author, Elizabeth Berry.
 p. ; cm. -- (Series in medical physics and biomedical
engineering)
 Includes bibliographical references and index.
 ISBN 978-1-58488-824-6 (hardback : alk. paper) 1. Imaging systems in medicine. 2.
Image analysis--Data processing. 3. Image processing--Digital techniques. I. Title. II.
Series.
 [DNLM: 1. Image Processing, Computer-Assisted--methods. 2. Diagnostic
Imaging--methods. WN 26.5 B534p 2007]

 R857.O6B47 2007
 616.07'54--dc22                                                    2007027399

**Visit the Taylor & Francis Web site at**
**http://www.taylorandfrancis.com**

**and the CRC Press Web site at**
**http://www.crcpress.com**

*To Jemima W. Berry*

# Contents

# *Preface*

In spite of the highly visual nature of image processing, and the ready availability of software tools, most image processing textbooks approach the subject either from a mathematical perspective or using a computer science approach. *A Practical Approach to Medical Image Processing* has been written in acknowledgment that digital image processing tools are now widely available, and that there are groups for whom neither traditional approach is suitable. This is especially true in medical imaging, where students in higher education and individuals in a variety of professional groups want to work with images and perform image processing, but without programming. The approach is very practical and does not assume a strong mathematical background when describing the underlying principles. There are activities throughout the book that use the freely available software package ImageJ. Using software helps to reinforce didactic points, and by learning to use a complete image processing package (rather than something written especially for teaching purposes) readers will be in a strong position to apply their skills in the future.

This textbook is ideal for taught postgraduate courses that involve image processing, such as those in radiography, medical imaging, medical physics and biomedical engineering. The book is very suitable for distance learning with feedback included for all questions. The book is also applicable for undergraduate use, and would provide a good introduction to medical image processing for computing students beginning a medical imaging project, for example. It will also be of interest to practitioners in the range of medical specialties, from radiology to dermatology, in which imaging plays an important role.

The book concentrates on image processing and does not attempt to discuss the principles of medical imaging, which are more than adequately covered by other books. The medical images are treated in a generic way to demonstrate the features that all digital images have in common, and examples come from a range of imaging modalities. This approach emphasizes the wide applicability of the image processing methods, and means that the book could be used by those wishing to apply image processing in fields other than medicine.

The first nine chapters introduce the main topics in image processing, and the reader will become increasingly proficient in using ImageJ as he or she progresses through the book. To offset the possibility that this highly practical way of learning could encourage an ad hoc approach to image processing, relevant points about good practice are highlighted at the end of each chapter. The issue of good practice is covered as a topic in itself in Chapter 10. Five case studies, each based on a published journal article, are presented in Chapter 11. The case studies demonstrate how the image processing topics of the preceding chapters are used in real-life studies. The final chapter contains suggestions on extensions to the material that will help instructors to personalize and adapt their courses. The accompanying CD contains supporting material for the activities and case studies.

**Elizabeth Berry**

# *Acknowledgments*

Thank you to all those who gave me permission to reproduce or use their images, data, software and articles. I would like to express my gratitude to Odette Dewhurst and Sarah Bacon for their skills, efficiency and enthusiasm during the time we spent preparing one of the courses from which this book has grown. Many individuals, especially at the University of Leeds and the Leeds Teaching Hospitals NHS Trust, have added to my knowledge and generously shared their expertise over the years. Finally, I would like to thank Wayne Rasband, the author of ImageJ, because without ImageJ this book could not have been written.

# On the CD

The CD contains software, digital images and documents to be used in some of the practical activities. It is organized into folders; there is one for each chapter and one containing software. The main piece of software on the CD is the Windows®* version of ImageJ. [1–3] The most recent update for Windows, and versions for other operating systems, can be obtained from the ImageJ website. The author of ImageJ is Wayne Rasband (Research Services Branch, National Institute of Mental Health, Bethesda, Maryland).

The disclaimer for ImageJ reads as follows: ImageJ is being developed at the National Institutes of Health (NIH) by an employee of the federal government in the course of his official duties. Pursuant to Title 17, Section 105 of the United States code, this software is not subject to copyright protection and is in the public domain. ImageJ is an experimental system. NIH assumes no responsibility whatsoever for its use by other parties, and makes no guarantees, expressed or implied, about its quality, reliability, or any other characteristic.

The disclaimers associated with the software on the CD are in the Readme-ForSoftware.txt file on the CD, together with further information. The ReadmeForImages.txt file contains information about the image data on the CD.

### Instructions for installing ImageJ:

The following are the Windows installation instructions for ImageJ 1.37 with extra plugins, which is supplied on the CD accompanying this book. To install ImageJ on your computer:

- Insert the CD in the drive and click on the Windows Start button. Select Run and then use the Browse button to navigate to the Software folder on the CD.
- Select the program ij137-jdk15-setup.exe, click on Open and then OK.
- Follow the instructions with the following guidance:
  - Install ImageJ in C:\Program Files\ImageJ (which is the default location) if possible. If you choose to put it elsewhere, make a note of the path chosen.
  - Accept the suggestion of ImageJ as the Start menu folder.
  - Add a ✓ to create a desktop icon. If you also want a Quick Launch icon, add a ✓ there, too.

---

* Registered trademark of Microsoft Corporation, Redmond, Washington, USA.

- Wait for the installation to take place.
- Remove the ✓ (by clicking on it) beside "Launch ImageJ" and then click on the Finish button.

To install the extra plugins required for following the activities in this book:

- Close ImageJ if it is running, using the cross at the top right.
- Use Windows Explorer to view the contents of the CD. Find the folder Software/ExtraPlugins and open it. You should see two folders here: BIJplugins and TurboReg, and two further files.
- Use "Edit, Select all" and then "Edit, Copy" to copy everything in the ExtraPlugins folder. Windows Vista users will find Select all and Copy in the Organize menu, or may click the Alt key on their keyboard to display a toolbar showing the Edit menu.
- Navigate in Windows Explorer to find the folder in which ImageJ was installed (C:\Program Files\ImageJ or the alternative chosen).
- Find the folder "plugins" and open it. There should already be several folders here; Analyze, Demos, Filters, Graphics, Input-Output, Macros, Stacks and Utilities. Use "Edit, Paste" to paste the folders and files copied a moment ago into this plugins folder.

When ImageJ is run, it will perform some preliminary installation tasks. It may be necessary to set the computer's firewall so that Java®* can operate.

---

## References

1. URL to download ImageJ: http://rsb.info.nih.gov/ij/download.html.
2. Rasband, W.S. ImageJ, U. S. National Institutes of Health, Bethesda, Maryland, USA, http://rsb.info.nih.gov/ij/, 1997–2006.
3. Abramoff, M.D., Magelhaes, P.J., and Ram, S.J. Image processing with ImageJ, *Biophotonics International*, 11, 36, 2004.

---

* Java and all Java-based marks are trademarks or registered trademarks of Sun Microsystems, Inc. Santa Clara, California, in the United States and other countries.

### E-books

Please go to http://www.crcpress.com/product/isbn/9781584888246 and click on the "Downloads/Updates" tab to obtain the material that is supplied on a CD with the printed book. Download the zip file and extract the contents to a folder. There should be 13 folders and a file called ReadmeForImages.txt. When the CD is mentioned in the e-book, use the relevant files from your folder instead of the CD.

# 1

## Image Processing Basics

## 1.1 Introduction

In this introductory chapter the basic terminology of image processing is presented. The ImageJ program is introduced using practical exercises, so that the reader is able to perform image processing tasks. Fundamental imaging processing principles, covering both the grayscale and spatial properties of an image, are described. The chapter closes with an overview of the content of the book, which is based around the five classes of image processing: image enhancement, image restoration, image analysis, image compression and image synthesis.

### 1.1.1 Learning Objectives

When you have completed this chapter, you should be able to

- Define the important basic terms used in image processing
- Understand methods of image enhancement that use the histogram
- List several different measurements, both spatial and grayscale, which can be made from images
- Discuss the five classes of image processing
- Use ImageJ to load and save an image, and perform simple tasks

### 1.1.2 Example of an Image Used in This Chapter

The first self-assessment question in each chapter is based on one of the images that appears later in the chapter. The question usually requires some background knowledge of medical imaging technology and is intended to help those who have previously studied medical imaging to start to integrate medical image processing with their existing knowledge. Those who lack this background knowledge should not be concerned if they cannot answer the first question in each chapter, and note that no prior knowledge is assumed in the other questions in each chapter.

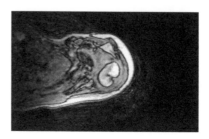

**FIGURE 1.1**
Image of a shoulder for self-assessment question 1.01.

*Self-assessment question 1.01*

Figure 1.1 is an example of an image used in this chapter. It shows the shoulder joint. The original image was 256 × 160 pixels in size and each pixel represented a 0.8 mm cube of tissue. Which imaging modality has been used to acquire the image shown? Note the loss of image information towards the left of the image. What does this suggest about the particular equipment that was used to acquire the image?

## 1.2   Definition of Image Processing

Image processing refers to the manipulation and analysis of pictorial information, for example by improving, correcting, changing or analyzing the image. The simplest example is the action of prescription spectacle lenses, which are designed to counteract distortion caused by defects in the owner's eye. The action of the lens is called an *optical process* because the lens is acting on an optical image. Optical images can be converted to analog images. An analog image is a two-dimensional representation of a scene; the image contains continuously varying tone (e.g., in a monochrome or black and white photograph) or continuously varying color.

### 1.2.1   The Digital Image

An analog image can be sampled to give an array of discrete points, each of which has an associated brightness value. This is a *digital image*, which is simply an array of numbers, one for each sample point (Figure 1.2). Each sample point is known as a picture element, or *pixel*. An image is generally sampled using a rectangular array of pixels. The process of sampling and assigning the numerical value is called *digitization*. Thus a digital image is an array of pixels, each of which has a value. When these data are displayed in the correct arrangement, with each pixel value used to give the brightness or color at that point, we see the image. The image *size*, or *matrix*, is expressed

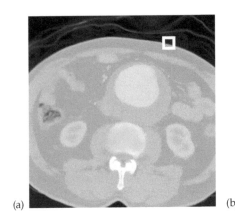

**FIGURE 1.2**
A digital image is an array of pixels, each of which has a value. (a) A digital image showing the location of a small square region of 15 × 15 pixels. (b) An enlargement of the 15 × 15 region of the image. The numerical values associated with the central pixels are shown overlaid.

in terms of the number of pixels in the rectangular array, for example 256 × 256 pixels. When image processing is performed on such an image, it is more fully described as digital image processing. In medicine, some digital images are acquired in digital form from the outset (e.g., computed tomography), while others are acquired as analog images (e.g., a chest x-ray on film) that may later be digitized. Three-dimensional data sets consist of a series of images each representing a slice through the object. In this case pixels are known as voxels.

## 1.2.2　Image Resolution

The *resolution* of an image is a measure of the fidelity (faithfulness) of the representation of the original scene. Resolution is related firstly to the characteristics of the imaging system and secondly to the number of pixels and to the range of brightness values that are used for digitization. In the discussion in this chapter, it will be assumed that the digitization process is performed on a perfect representation of a scene, so the properties of the imaging system may be ignored. The effects of the imaging system are considered in Chapter 7.

The question addressed when considering image resolution is "Does the digital image resolve elements of the original scene?" There are two components of resolution: grayscale resolution and spatial resolution. The question therefore has two parts: Can objects that are similar in tone or color be distinguished? Can small objects be seen?

### 1.2.2.1　*Grayscale Resolution*

The digitization process assigns a number to each pixel to represent the brightness at that location in the original image. The grayscale, or brightness, (a)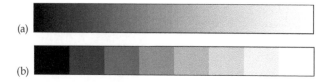

(b)

**FIGURE 1.3**
Grayscales associated with images of two different bit depths. (a) An 8-bit grayscale (256 shades of gray). (b) A 3-bit grayscale (8 shades of gray).

resolution is affected by how many brightness values are allowed. It is very common to assign the pixel value to one of 256 discrete values (from 0 to 255). Each value is displayed as a separate shade of gray, and these shades run smoothly from dark (0) to light (255). An image with 256 discrete values is known as an 8-bit image, because $2^8 = 256$. An 8-bit grayscale is shown in Figure 1.3a.

It is quite possible to digitize to gray scales using a different number of shades of gray. For example, 128 shades of gray are associated with a 7-bit image (as $2^7$ is 128), and 8 shades of gray (Figure 1.3b) gives a 3-bit image, since $2^3$ is 8. In Figure 1.4a through g, one image is shown displayed with several different grayscale resolutions. Notice that with decreasing grayscale resolution (i.e., fewer gray values) the images appear progressively coarser. *Contouring* occurs when too few gray levels have been used; gradual changes in image brightness are rendered as discrete bands.

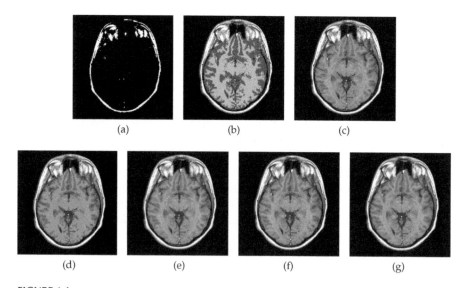

**FIGURE 1.4**
An image displayed using a range of grayscale resolutions. (a) 2 gray levels (1-bit), (b) 4 gray levels (2-bit), (c) 8 gray levels (3-bit), (d) 16 gray levels (4-bit), (e) 64 gray levels (6-bit), (f) 128 gray levels (7-bit), (g) 256 gray levels (8-bit).

The *dynamic range* of a gray scale is the range of pixel values used, for example, from 0 to 255 for the full 8-bit scale.

*Self-assessment question 1.02*

How many shades of gray would there be in a 5-bit image scale?

*Self-assessment question 1.03*

What is the derivation of the word "pixel"?

### 1.2.2.2 Spatial Resolution

The digitization process assigns a number to each pixel to represent the brightness at that location in the original image. The spatial resolution is affected by how close together the samples are taken.

The concept of *spatial frequency* is useful when considering spatial resolution. Details in images are made up of changes from dark to light to dark. The rate of this change is called the spatial frequency. In Figure 1.5 the image has just one spatial frequency. In a real image, however, a range of spatial frequencies is present. The higher spatial frequencies correspond with fine detail, and low spatial frequencies with larger objects that have little change of gray level within. Sampling theory requires that the image must be sampled so that at least two samples are acquired within the smallest detail that is to be represented. Thus, the highest spatial frequencies that can be resolved in a digital image must have a frequency half that of the sampling rate. If sampling is done at a low rate, high spatial frequency details will not be seen in the digital image. This effect is demonstrated in Figure 1.6. The sampling rate gets progressively lower from Figure 1.6a to e, leading to low spatial resolution, and a consequent lack of detail. The effect where the pixels are very apparent is called *pixelation*.

Spatial aliasing occurs when an image has been undersampled (Figure 1.7). The term aliasing is derived from the word "alias." High spatial frequencies are misrepresented as lower ones, and so could be considered to have

**FIGURE 1.5**

An image with a single spatial frequency associated with the frequency of the sinusoidal variation in image intensity. Typical medical images include a range of spatial frequencies ensuring that features of different scales are represented.

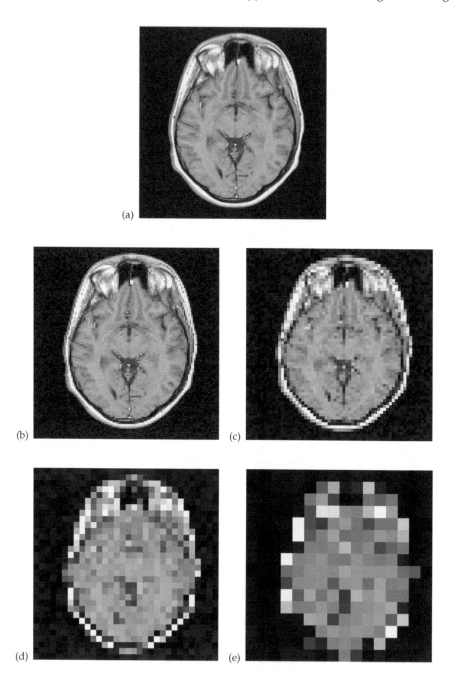

**FIGURE 1.6**
The effect of sampling rate on image resolution. (a) Original, (b)–(e) the effect of repeatedly halving the sampling rate.

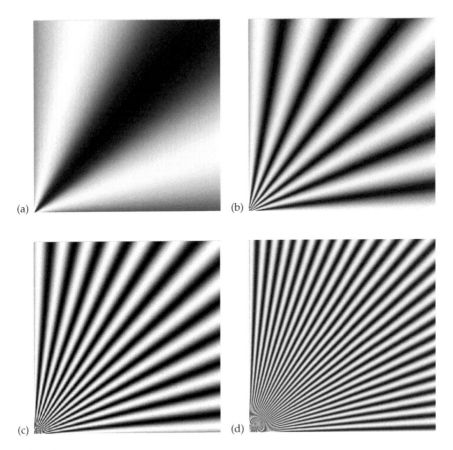

**FIGURE 1.7**
Sampling rate and aliasing. In each image, spatial frequency increases towards the bottom left hand corner. The spatial frequencies present in each image increase from (a)–(d). All the images have been sampled at the same rate. There is no aliasing in (a), which contains low frequencies. In (b)–(d), Moiré patterns appear when the spatial frequency is over half the sampling rate, leading to the occurrence of apparent low-frequency features that are not in fact present in the original. Aliasing requires repetitive high-frequency details, so is not often seen in clinical images, but may be seen in images of test objects.

assumed an alias. In practice, aliasing is not often seen, unless an image has repetitive high-frequency details.

Spatial resolution can be expressed numerically in several different ways. One of the most common is to determine the maximum number of black and white stripes in 1 mm that can be resolved by the eye, and do not blur into gray. This is quoted as a number of line pairs per millimeter, or lpmm$^{-1}$. To determine spatial resolution from an image, it is obviously necessary to know the pixel size in millimeters. This can be achieved by calibration, which is described in Chapter 5.

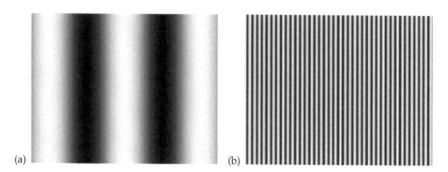

**FIGURE 1.8**
(a)–(b) Images for self-assessment question 1.04.

*Self-assessment question 1.04*

Which of the patterns in Figure 1.8 has the higher spatial frequency?

### 1.2.3  Image Data Types

Computer operations are based on binary arithmetic, where there are just two possible options, 0 or 1. In imaging, an image with only two possible gray values is termed a binary image. Such an image can be displayed using a 1-bit scale because there are two possible options requiring $2^1$ bits. Binary images, however, for reasons of compatibility, are often stored using a data type that would allow many more gray values.

Integer numbers are most easily stored in a computer in either 1-byte (8 bits) or 2-byte (16 bits) format, so it is most common for images to be 8-bit or 16-bit images. A 16-bit image corresponds with a gray scale having $2^{16}$ gray levels running from 0 to 65535. It is rare for medical images to have such a wide dynamic range. For example, in MRI a maximum value of 4095 is very common, corresponding with a 4096 gray level scale, which corresponds with 12 bits. However, because of the ease of storage in 2 bytes, the 16-bit format will often be used even if the actual image gray values only occupy the lower end of the scale.

The image data types discussed so far are unsigned types, allowing storage of values of 0 and above. Signed data types can represent the same number of gray values as their unsigned counterpart, but the range of values will differ. For example, 256 values can be stored using an 8-bit data type. The range of values stored in an unsigned 8-bit image is from 0 to 255; in a signed 8-bit image it will be –128 to 127.

*Self-assessment question 1.05*

This is a list of some of the image data types recognized by various image processing and analysis programs (for example, ImageJ): binary, 8-bit unsigned, 8-bit signed, 16-bit unsigned, 24-bit color and floating point.

- How many gray values are expected for the binary data type?
- Do the signed and unsigned 8-bit data types both support the same maximum value?
- Why are 16 bits often stored even when only 12 bits are used?

It is worth commenting on the 24-bit color data type. This is an interesting representation, where each pixel has three different 8-bit scales associated with it, one each for red, green and blue. It is sometimes known as an RGB scale. Each of the three scales runs from 0 to 255, and the color in which the pixel is displayed depends on the value in all three. For example, if all three are 255 then the pixel is white, but if, for example, the value was 0 for the green channel, and 255 for the red and the blue, the result would be magenta.

*Self-assessment question 1.06*

Select the binary image from the four images in Figure 1.9.

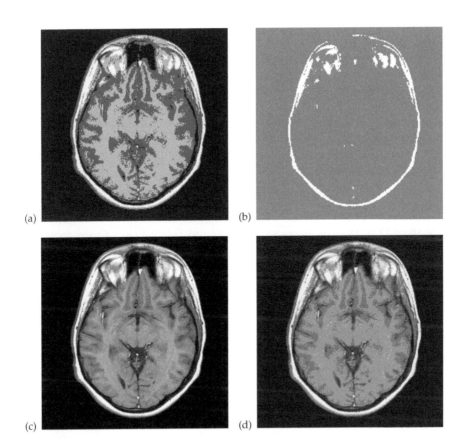

(a)

(b)

(c)

(d)

**FIGURE 1.9**
(a)–(d) Images for self-assessment question 1.06.

*Self-assessment question 1.07*

Define what is meant by the dynamic range of a gray scale.

## 1.3 Introduction to ImageJ

In the rest of the book, the program ImageJ will be used extensively to support material in each chapter. In this section, basic skills are covered, such as loading images, so that in later activities familiarity can be assumed.

### Activity: Starting and Closing ImageJ

Start ImageJ. Depending on how your system is set up

1. Use the microscope icon on your desktop, or
2. Start | Programs | ImageJ | ImageJ.

The program will appear on your desktop: the initial display is one window, with a dropdown menu row and a row of buttons. You will still see your wallpaper and desktop icons, as in the example in Figure 1.10. Try dragging the small window around your monitor.

Leave ImageJ running at present, but note that when you have finished using ImageJ, the program may be closed if you select File | Quit. If there are any images open that have not been saved following changes, you will be asked if you wish to save them.

### Activity: Loading an Image with ImageJ

Place the CD in the computer's CD drive and select File | Open from the main ImageJ menu (Figure 1.11).

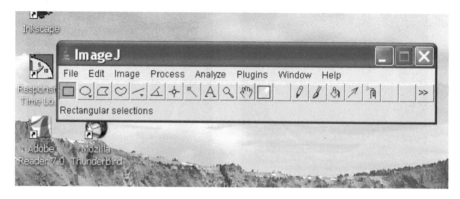

**FIGURE 1.10**
The ImageJ window on a Windows desktop.

**FIGURE 1.11**
File | Open on the main ImageJ menu.

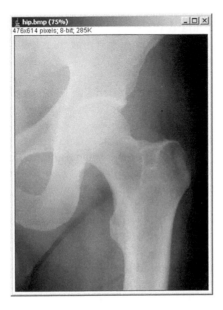

**FIGURE 1.12**
The image hip.bmp when opened using ImageJ. Information about the image data size and type is shown in the panel above the image.

ImageJ displays a standard Windows®* "Open File" window. Navigate to the CD drive, select the folder for Chapter01 and then the file hip.bmp. Click on the "Open" button.

The image will be displayed in its own window on your desktop, with information about its size and data type in the white panel (Figure 1.12).

Run the cursor across the image (without clicking). At the bottom left of the ImageJ menu, the status bar changes to show the $x$ and $y$ coordinates of the cursor, and the gray value of the pixel at that location. Notice that all the pixel values are integers in the range from 0 to 255 as is expected for an 8-bit image.

---

* Registered trademark of Microsoft Corporation, Redmond, Washington, USA.

Select File | Quit to close ImageJ. You will be asked if you wish to save any open images; answer No.

### Activity: 16-Bit and 8-Bit Images

Start ImageJ.

Place the CD in the computer's CD drive and select File | Open from the main ImageJ menu (Figure 1.11).

ImageJ displays a standard Windows "Open File" window. Navigate to the CD drive, select the folder for Chapter01 and then the file chapter01.tif. Click on the "Open" button.

The image is a 256 × 256 pixel image, in the 16-bit data type, and this information is displayed in the white panel above the image.

Run the cursor across the image (without clicking), and while doing so check the status bar at the bottom left of the ImageJ menu. Note how in this 16-bit image there are pixel values much greater than the maximum of 255 that could be accommodated by the 8-bit data type.

Select Image | Type | 8-bit; this will convert the data type of the image from 16-bit to 8-bit.

Note that the data type indicated in the white panel above the image is now 8-bit. When the cursor is run across the image, the pixel values are now all in the 8-bit range from 0 to 255. The image looks much the same, because the image display uses only values between 0 and 255 for displaying both 16-bit and 8-bit images. However, the dynamic range of the actual gray values in this image has been much reduced by the reduction from 16 to 8 bits.

Click on the cross at the top right of the image window to close the image. Answer No when you are asked if you wish to save the changes to the image.

*Self-assessment question 1.08*

Variations of the image chapter01.tif are shown in Figure 1.13. By eye, place the images in descending order of spatial resolution.

### Activity: Spatial Resolution

Start ImageJ if it is not already running.

Place the CD in the computer's CD drive and select File | Open from the main ImageJ menu. Open the file chapter01.tif and then open chapter01_c.tif.

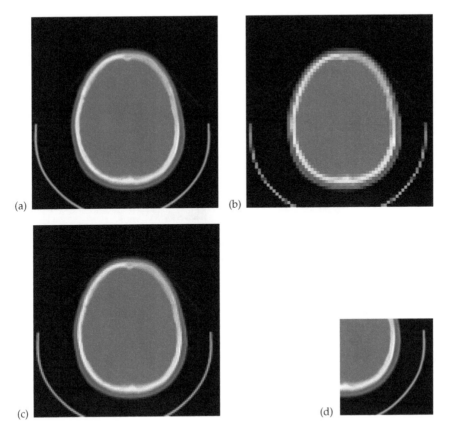

**FIGURE 1.13**
Images for self-assessment question 1.08. (a) 256 × 256 pixels, (b) 64 × 64 pixels, (c) 128 × 128 pixels, (d) 128 × 128 pixels.

By viewing the information in the panel above each image, it can be seen that chapter01.tif is a 256 × 256 pixel 16-bit image, and chapter01_c.tif is a 128 × 128 pixel 8-bit image. However, the two images both have the same spatial resolution, because the same number of pixels represents the same object distance in each case.

Select File | Quit to close ImageJ. You will be asked if you wish to save any open images; answer No.

## 1.4 Grayscale Image Processing Basics

### 1.4.1 Color Look-Up Tables (LUT)

In medical imaging it is common to apply false color to an image while retaining the original 8-bit scale. Instead of assigning a shade of gray to each

**FIGURE 1.14 (Please see color insert following page 16)**
The thermal color look-up table, which is a smoothly varying scale running from black (0) to white (255) through shades of red, orange and yellow.

**FIGURE 1.15 (Please see color insert following page 16)**
The 5-ramps color look-up table, which is a discontinuous scale with black representing five different gray values.

pixel value, a color is assigned. Some color scales are continuous scales, such as the thermal scale shown in Figure 1.14 where, just as with the gray scale, there is a smooth graduation in shade. Other scales change color abruptly, for example, with the value 101 being green and 102 being black (Figure 1.15). Abrupt changes of this kind can lead to the appearance of a sharp boundary in an image where in fact no physical feature is present. Color LUTs add impact to an image, but care is required to avoid misleading the viewer.

### Activity: Look-Up Tables in ImageJ

Start ImageJ, or close any open images (by clicking on the cross at the top right of the image window) if ImageJ is already running.

Select Image | Lookup Tables > and then choose a table from the list given, for example, Fire or Spectrum.

A 256 × 32 ramp image will be created to display the LUT. An 8-bit ramp image has pixel values from 0 to 255 placed in ascending order across the image.

If images are already open in ImageJ, the LUT is applied to the currently selected image. ImageJ allows you to have up to 1000 images open at a time, so you need to click on the image for which the LUT is required before selecting Image | Lookup Tables > .

To return the image display to its original appearance, select Image | Lookup Tables > Grays.

Sometimes the sequence of LUTs selected may result in the image appearing to have an inverted gray scale at this point (what was previously white is displayed in black). If this happens, select Edit | Invert. Use the Invert option any time if you would prefer the gray scale to be inverted.

Close any open images and Select File | Quit to close ImageJ.

### 1.4.2 Image Contrast

In an image, it is possible to recognize an individual structure because it differs in brightness from its surroundings. This degree of visibility, representing the difference between light and dark regions of the image, is called the image contrast. If an image has low contrast, it means that it is difficult to distinguish different structures. The contrast in an image is affected by the image acquisition mechanism itself, as contrast between different materials relies on there being a different amount of interaction between the imaging radiation and the different materials.

*Self-assessment question 1.09*

In Figure 1.16, which image has the lower contrast?

There is a related quantitative parameter called *contrast*. For the two regions *S* and *B* illustrated in Figure 1.17, contrast is defined as

$$C = \frac{\left(S_S - S_B\right)}{S_B} \qquad (1.1)$$

where $S_S$ represents the mean signal within the region and $S_B$ is the mean signal in the background. Contrast may also be expressed as a percentage by multiplying the result by 100. Be aware that other definitions are sometimes used, for example, a simple difference in signal.

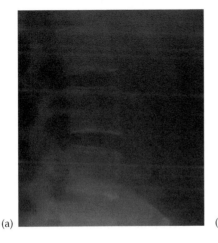
(a)

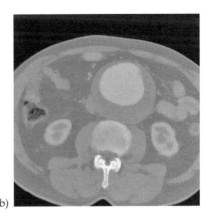
(b)

**FIGURE 1.16**
(a)–(b) Images for self-assessment question 1.09

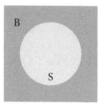

**FIGURE 1.17**
For the two regions $S$ and $B$, contrast is defined as $C = (S_S - S_B)/S_B$ where $S_S$ represents the mean signal within the region and $S_B$ is the mean signal in the background.

### Contrast to Noise Ratio (CNR) and Signal to Noise Ratio (SNR)

The contrast to noise ratio (CNR) is a more useful definition than contrast alone when considering the visibility of features in a noisy, real-life image. CNR is found by dividing the measured contrast by the background standard deviation. The background standard deviation is measured in an area of the image where the object is not present, such as the image background. In a similar way, the signal to noise ratio (SNR) is given by the mean signal divided by the measured standard deviation of the image background.

## 1.4.3 The Image Histogram

The grayscale image histogram is a way of illustrating the distribution of gray levels in an image. It shows how many pixels have particular gray values. The information may be displayed as a graph, or presented as a table of numbers. In the graphical representation, the horizontal axis is used to indicate the grayscale values, which may be shown individually or in ranges, and the vertical axis shows how many pixels have each gray value. An example is shown in Figure 1.18. Once the histogram has been calculated, the underlying data may be adjusted in order to perform image enhancement. The histogram is known as a *global* histogram if it is calculated for the whole of the image. Alternatively, the histogram found using only part of the image, or *region of interest*, is called a *local* one.

### 1.4.3.1 Image Contrast and the Histogram

Three commonly seen types of image are those with low contrast, high contrast or with well-balanced contrast. In a typical low-contrast image (Figure 1.19a), different features in the image all have much the same gray level, so that only a few of the full range of gray levels are actually used, and these are close together. This is what is seen in a low-contrast histogram (Figure 1.19b): the pixel values are closely grouped together, and the other gray values are unoccupied or have very few pixels associated with them.

**FIGURE 1.14**
The thermal color look-up table, which is a smoothly varying scale running from black (0) to white (255) through shades of red, orange and yellow.

**FIGURE 1.15**
The 5-ramps color look-up table, which is a discontinuous scale with black representing five different gray values.

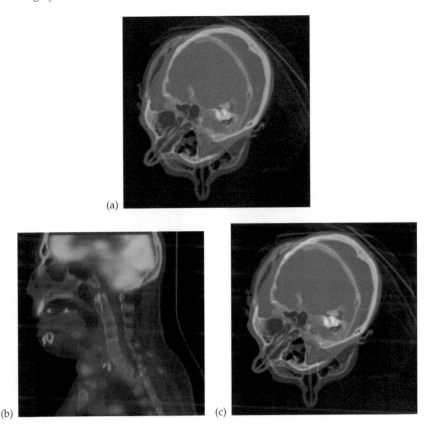

**FIGURE 8.17**
Pseudocolor displays. (a) Two unregistered slices using red and green only, (b) registered PET and CT images using red and green only, (c) two unregistered slices using red, green and blue. (Image data from the Chapel Hill Volume Rendering Test Data Set, Volume I. [8] and http:// pubimage.hcuge.ch:8080/. With permission.)

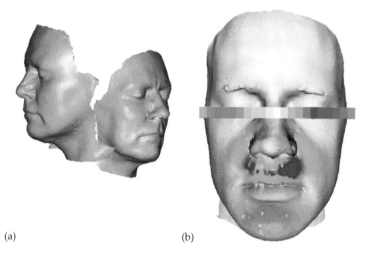

(a)                                            (b)

**FIGURE 8.19**
(a) Two surface laser scans acquired at different times. (b) Image calculated, for a different individual, by registering surface laser scans taken pre- and postoperatively. The color scale has 2 mm increments. The red end of the scale indicates movement backwards as a result of surgery, and the purple indicates movement forwards as a result of surgery. (From Miller, L., Morris, D.O., and Berry, E. Visualizing three-dimensional facial soft tissue changes following orthognathic surgery, *Eur. J. Orthod.*, 29, 24, 2007. With permission of Oxford University Press.)

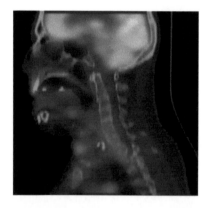

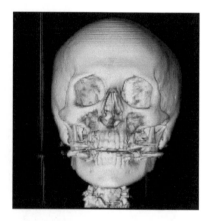

**FIGURE 10.2**
The appearance of Figure 8.17b to a viewer with the severe red–green color deficiency deuteranopia. (This image was generated using software from www.vischeck.com. Image data from http://pubimage.hcuge. ch:8080/. With permission.)

**FIGURE 11.24**
Result of rendering a stereo pair and generating a red-green anaglyph. (Image data from the Chapel Hill Volume Rendering Test Data Set, Volume I. [8] With permission.)

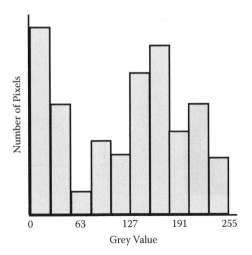

**FIGURE 1.18**

The graphical representation of an image histogram. The horizontal axis indicates the gray-scale values, and the vertical axis indicates the number of pixels with each gray value.

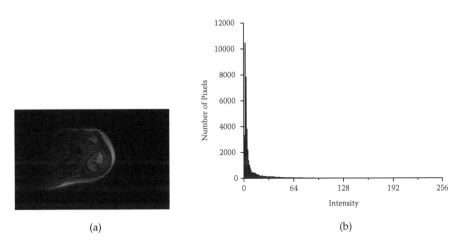

(a)                                                                                      (b)

**FIGURE 1.19**

Low image contrast and the histogram. (a) A low-contrast image, (b) the associated image histogram.

On the other hand, high-contrast images have gray levels that fall into separate, very different, groups. An extreme case is shown in Figure 1.20a, where the image has pixels with only two gray levels, 0 and 255. High-contrast histograms are *bimodal*, which means that there are two well-separated peaks, with few values occupied in between (Figure 1.20b).

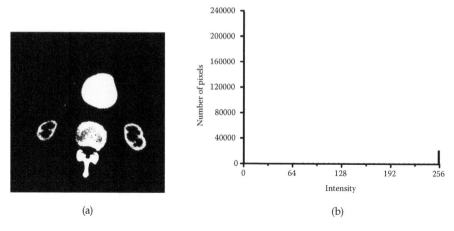

(a)                                          (b)

**FIGURE 1.20**
High image contrast and the histogram. (a) A high-contrast image, (b) the associated image histogram; note that many pixels have an intensity value of zero.

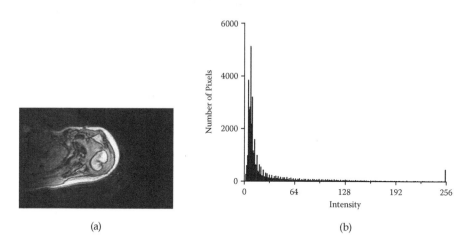

(a)                                          (b)

**FIGURE 1.21**
Well-balanced image contrast and the histogram. (a) An image with balanced contrast, (b) the associated image histogram. Note the high occupancy of the lowest gray values, which are associated with image background.

In an image with well-balanced contrast, gray levels throughout the range are occupied (Figure 1.21a). This property can be seen in the histogram, but care needs to be taken when considering medical images, because these often have a large number of black background pixels. The background pixels have low gray values, sometimes but not always with value 0. The eye will automatically exclude the background from its assessment of the image

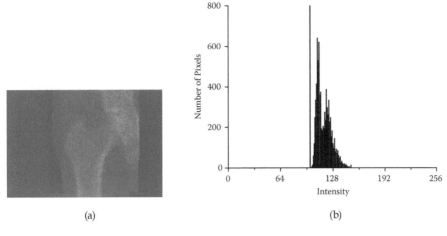

(a)                                        (b)

**FIGURE 1.22**
Although the dynamic range of an 8-bit image is 256, an 8-bit image can have a smaller actual dynamic range. (a) An image with an actual dynamic range of 50, from 10 to 150, (b) the associated image histogram.

contrast, but the background pixels will be included in any analyses by a software tool unless specifically excluded. The maximum value on the vertical axis corresponds with the number of pixels with the most popular gray value, and if there is a large number of background pixels in an image, this number may be many more times the number with any other gray value. The result is that the scale of the vertical axis is such that it is very hard to see the occupancy of the gray values other than for the background regions. This effect is shown in Figure 1.21b.

### 1.4.3.2 Actual Dynamic Range

The available dynamic range for an 8-bit image is from 0 to 255. The image histogram may be used to see how much of that range is actually occupied. For example, in Figure 1.22 the image has a small *actual dynamic range*, and there is a peak in the histogram, with all the pixels having values between 100 and 150. This is related to the earlier discussion about image contrast. An image with a small dynamic range will have low contrast and poor brightness resolution.

*Self-assessment question 1.10*

In the histograms in Figure 1.23, gray level 0 represents black and 255 represents white. Assign the correct description from the following list to each histogram: a predominantly light image, an image with a wide dynamic range, a well-balanced image with a large dynamic range, an image with a small dynamic range, a high-contrast image with a large dynamic range and a predominantly dark image.

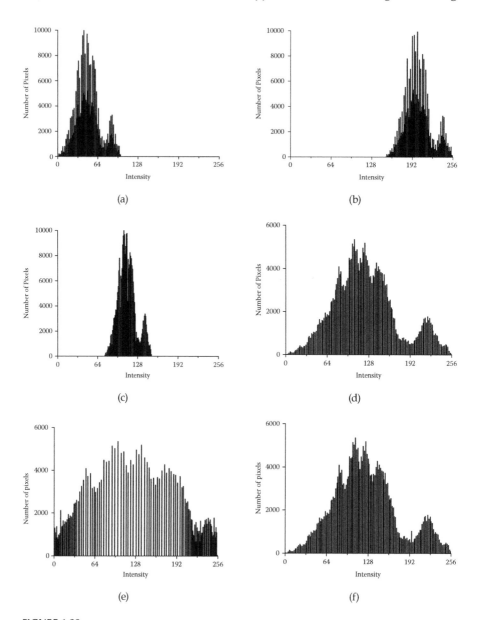

**FIGURE 1.23**
(a)–(f) Histograms for self-assessment question 1.10.

### 1.4.3.3   *Histogram Operations*

The contrast properties of an image may be changed by changing the pixel values in a way that has a particular effect on the histogram. The aim of *histogram equalization,* for example, is to spread pixel values evenly across the whole range, so that there are approximately equal numbers of pixels with each gray value.

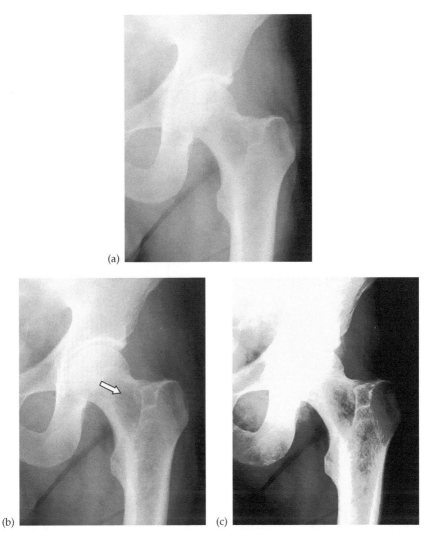

**FIGURE 1.24**
Histogram equalization. (a) Original, (b) result of histogram equalization [note the enhanced image contrast in the arrowed area], (c) result of histogram equalization based on a region of interest in the femoral neck.

The effects of histogram equalization can be seen in Figure 1.24. Figure 1.24a is the original image and Figure 1.24b shows the result of histogram equalization. The enhanced image contrast is noticeable, particularly in the femoral neck (arrowed). The associated image histograms are in Figure 1.25. The equalization effect (sometimes called flattening) is evident in the reduction in the height of the peaks and the filling of the troughs in the histograms. The improved contrast in the femoral neck arises because the pixels in this area cover a greater range of values after processing than they did before.

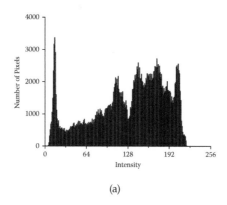

(a)

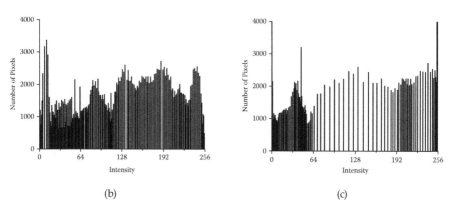

(b)                                                        (c)

**FIGURE 1.25**
Image histograms associated with the images in Figure 1.24. (a) Original, (b) after histogram equalization, (c) histogram equalization based on a region of interest in the femoral neck.

Because of the global nature of the operation, not all areas of the image will have greater contrast after processing. If the area of interest is known, it is possible to perform a local operation, where the equalization is based on the histogram in the selected area (Figure 1.24c and Figure 1.25c). Most pixels outside the region of interest have been set to values of 0 and 255 at the extremes of the range, and those in the region of interest are distributed evenly across the whole range.

### Activity: Histogram Equalization in ImageJ

Start ImageJ and open a copy of hip.bmp. Open a second copy of hip.bmp.

Select one of the two images by clicking on the image, then select Analyze | Histogram. A histogram plot appears in its own little window (Figure 1.26). The window may be moved around the monitor by clicking and dragging.

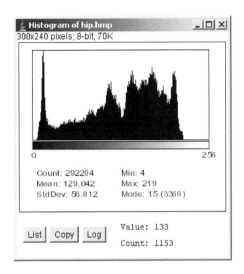

**FIGURE 1.26**
The ImageJ histogram window.

Make sure that one of the images is selected by clicking on it. The bar across the top will be blue, not gray, when the image is selected.

Select Process | Enhance Contrast. In the dialog box add a tick next to Equalize Histogram, then, while holding down the Alt key, click on OK. This displays the result of histogram equalization on the image.

Do not check to see how the histogram has changed; instead, this is a good time to try out the Undo function. Select Edit | Undo. The image should revert to its previous appearance, but it may not if any other processing was performed after the histogram equalization. Repeat the histogram equalization. Visually compare the appearance of the result with the unprocessed image.

Select Analyze | Histogram to get a window similar to the one in Figure 1.27. Note the difference between the original histogram and this new, equalized one.

If the histogram equalization operation doesn't appear to be making any difference to the image, try closing the images by clicking on the cross at the top right of the image windows, and start the activity again.

### More Histogram Equalization

Look in the Process Menu section of the ImageJ Documentation to see which operation is performed if the Alt key is not held down when performing histogram equalization in ImageJ. Compare the results obtained with and without use of the Alt key.

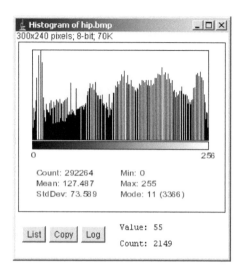

**FIGURE 1.27**

The ImageJ histogram window for an image after histogram equalization.

## Activity: Saving an Image with ImageJ

Click on the image to be saved; in this case it is the result of the histogram equalization.

Select File | Save As | BMP. Select a drive letter other than that of the CD drive, then type in the name of the file you wish to save. This should be different from the original file name, for example "hipequal.bmp". Click the "Save" button.

Check that the operation worked as follows. Close the image window in ImageJ by clicking on the cross at the top right of the image window. Load the new file hipequal.bmp, remembering that the new file is not on the CD.

Close hipequal.bmp and Select File | Quit to close ImageJ.

*Self-assessment question 1.11*

In the image in Figure 1.28, there are a large number of background pixels with values close to 0. Think about applying the process of histogram equalization, as described above, to such an image. Describe the image that will result from applying histogram equalization to the whole image, paying particular attention to the characteristics of the background of the image and the spread of the histogram. Suggest how the histogram equalization process might be adjusted to achieve a more satisfying image.

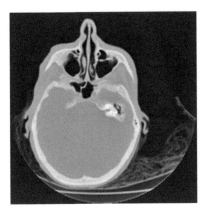

**FIGURE 1.28**
Image for self-assessment question 1.11. (Image data from the Chapel Hill Volume Rendering Test Data Set, Volume I. [1] With permission.)

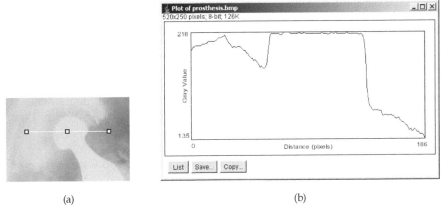

(a)　　　　　　　　　　　　　　　　　(b)

**FIGURE 1.29**
Profile statistics. (a) A straight line profile drawn across an image, (b) graphical representation of the pixel values along the profile.

## 1.4.4 Pixel Statistics

### 1.4.4.1 Profile Statistics

A profile across an image is shown in Figure 1.29a. The value of each pixel along the profile is shown on the plot Figure 1.29b, and may be saved to a file for further analysis. Image processing software, such as ImageJ, will often offer options for calculating statistics from the pixel values, such as the mean gray value along the profile.

**FIGURE 1.30**
The ImageJ straight line tool icon on the ImageJ Tools menu.

### Activity: Profile Statistics in ImageJ

Start ImageJ and load the image prosthesis.bmp.

Select the straight line tool icon in the ImageJ Tools menu (Figure 1.30) and click and drag on the image to draw a straight line (Figure 1.29a).

Select Analyze | Plot Profile. A profile similar to that in Figure 1.29b will be displayed.

See what happens when you click and drag on the little squares on the line.

Select Analyze | Plot Profile again to draw a new profile.

To export the values in the profile (for example, to analyze the profile outside ImageJ, or to use the data to create a plot suitable for publication), use the "Save" button in the plot window. Choose a folder and file name, using perhaps a .txt extension, and click on Save. If familiar with Excel®*, and with importing text files, try importing the profile into Excel now.

Select File | Quit to close ImageJ. You will be asked if you wish to save any open images.

#### 1.4.4.2  Region of Interest Statistics

A region of interest is an area (or a volume in 3-D) within which measurements of pixel statistics are made. The region may be drawn interactively, or could have been defined by previous image processing operations. The measured statistics might be incorporated into further processing (for example, a measurement of mean and standard deviation might be used as the basis for the choice of a threshold gray level) or may themselves be the values that are analyzed in a large study.

### Activity: Gray Level Measurements in ImageJ

Start ImageJ and open the image prosthesis.bmp

---

* Registered trademark of Microsoft Corporation, Redmond, Washington, USA.

**FIGURE 1.31**
The ImageJ rectangular selections icon in the ImageJ Tools menu.

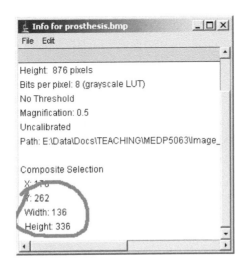

**FIGURE 1.32**
The ImageJ Info window.

Select the rectangular selections icon in the ImageJ Tools menu (Figure 1.31), and click and drag on the image to draw a box.

Select Image | Show Info. Look in the Info Window, where information on the width and height of the box is displayed (Figure 1.32). The selection box can be dragged around the image by clicking inside it. To delete the box, click somewhere else in the image. To resize the box, delete the box and start drawing again.

Select Analyze | Measure. Look in the Window entitled "Results," where statistics are displayed (Figure 1.33).

To include other statistics in the list of those to be measured, select Analyze | Set Measurements. A box appears listing all the measurements that can be requested (Figure 1.34). An obvious addition would be standard deviation. Ensure there is a tick beside the options you require, then click the "OK" button.

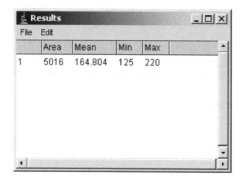

**FIGURE 1.33**
The ImageJ Results window.

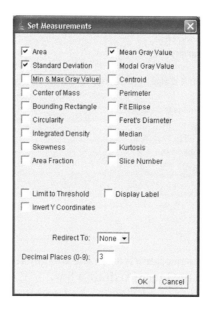

**FIGURE 1.34**
The ImageJ Set Measurements dialog.

Position a 20 × 20 square ROI in the bright region below the tip of the prosthesis in prosthesis.bmp. Watch the main ImageJ window to check on the size of the region while drawing. Measure the mean and standard deviation of the gray values in this ROI. The ROIs may not be exactly in the same place, so the results will probably be a little different from those shown in Figure 1.35.

Close any open images and Select File | Quit to close ImageJ.

*Self-assessment question 1.12*

Self-assessment question 1.11 was a thought experiment about the effect on histogram equalization of the presence of a large proportion of low pixel

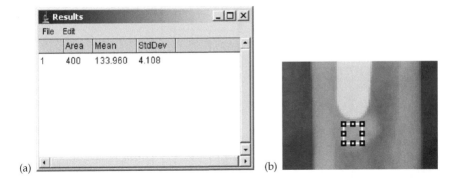

**FIGURE 1.35**
Measurement of statistics within a region of interest. (a) The ImageJ Results window, (b) the image selection for which the statistics were measured.

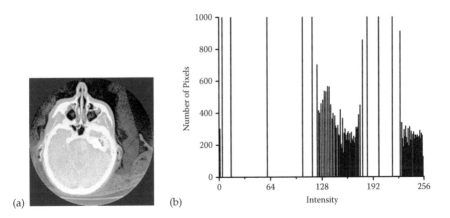

**FIGURE 1.36**
(a) and (b). Images for self-assessment question 1.12. (Image data from the Chapel Hill Volume Rendering Test Data Set, Volume I. [1] With permission.)

value background pixels. In ImageJ, if there is no region of interest defined, then histogram equalization is a global operation. However, if a region of interest has first been defined on the image, any subsequent histogram equalization will be based on the pixel values within that region. Load two copies of the image cthead.tif* from the CD and perform a global histogram equalization operation on one, and a local histogram equalization operation on the other. The aim is to achieve results similar to those in Figure 1.36 and Figure 1.37.

---

* Image data from the *Chapel Hill Volume Rendering Test Data Set*, Volume I. [1] With permission.

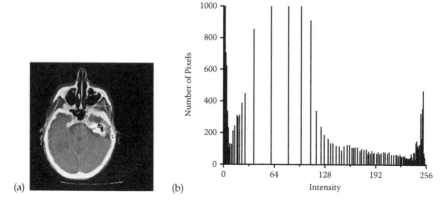

**FIGURE 1.37**
(a) and (b). Images for self-assessment question 1.12. (Image data from the Chapel Hill Volume Rendering Test Data Set, Volume I. [1] With permission.)

## 1.5   Spatial Image Processing Basics

### 1.5.1   Image Size

The size of an image is expressed in terms of its extent in pixels, for example, $512 \times 512$ pixels. The pixel extent is sometimes called the matrix size of an image. The file size in bytes of an image depends on the matrix size and the number of bytes stored for each pixel, as indicated in Section 1.2.3. When displaying an image, it is important to maintain the relationship between the width and height of the image, the *aspect ratio*, to avoid apparent distortion of the image.

### 1.5.2   Neighbors and Connections

When dealing with arrays of pixels, it is often useful to talk about the relationship of the pixels close to one another. There are conventions about how this is done. In Figure 1.38a the four pixels that abut the top, bottom, left and right of our pixel of interest (the central one) are called the *4-neighbors* of that pixel. They are also said to be *4-connected* to it.

*Self-assessment question 1.13*

Based on the definition of the 4-neighbors, suggest the term used to describe all the pixels other than the central one in Figure 1.38b.

### 1.5.3   Scaling and Rotation

The spatial properties of an image can be altered in a range of ways, the simplest of which are geometric operations to rescale or rotate the image. These

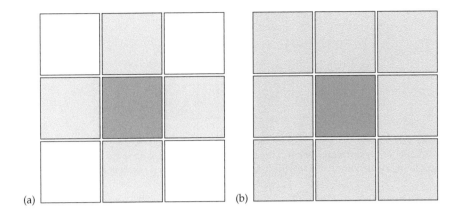

**FIGURE 1.38**
(a) The 4-neighbors of the dark gray central pixel shown in light gray. (b) Image for self-assessment question 1.13.

operations belong to a class called matrix transformations, because of the way in which the alteration is computed using a matrix.

### Activity: Image Rotation in ImageJ

Start ImageJ and open the image hip.bmp.

Select Image | Rotate | Rotate 90 Degrees Right.

Repeat, but select the Arbitrarily option and enter the number of degrees of rotation as required.

Close the image and exit from ImageJ.

### 1.5.4 Interpolation

Interpolation is an operation in which values are calculated for additional pixels that lie between the original pixels. The operation is performed if an image transformation means that new pixels are required in positions that were not represented in the original image. For three-dimensional data sets, interpolation is often performed to generate new slices between the original ones, so that the slice separation is the same as the pixel size within each slice. The cubic voxels that result are also described as being isotropic. The simplest method of interpolation, which is very widely used, is known as linear interpolation (bilinear in 2-D and trilinear in 3-D). The new pixel value is the distance-weighted average of the nearest gray values. In the example in Figure 1.39, the new pixels are midway between the original pixels, so the new values are simply the average of the pixel values on either side. Interpolation is not a way of increasing the spatial resolution of an image,

**FIGURE 1.39**

An example of linear interpolation. Three pixels are shown in the top row, and the result of interpolating two additional pixels midway between the original pixels is shown below.

because only the pixel size has changed, and there has been no change to the resolution at which the acquisition was made. It can, however, dramatically improve the appearance of images acquired using low resolution techniques, such as nuclear medicine.

## 1.6    The Five Classes of Image Processing

Digital image processing operations can be grouped into five fundamental classes:

- Image enhancement (e.g., contrast enhancement, spatial filtering, noise reduction)
- Image restoration (e.g., geometric correction)
- Image analysis (e.g., segmentation, classification)
- Image compression
- Image synthesis (e.g., registration, visualization, 3-D rendering)

An overview of these classes is given here. Operations from each class are discussed in detail in Chapters 2 to 9.

### 1.6.1    Image Enhancement

Image enhancement operations are used to improve some aspect of the quality of an image. Image enhancement is very often performed interactively by an observer aiming to increase the contrast or brightness of an image, sharpen details or remove noise. In this type of operation, the improvement is often subjective and dependent on the judgment of an observer, but quantitative measures of image quality do exist. Typically, the contrast to noise ratio of an image may be measured, or the receiver operating characteristic (ROC) generated to see if the image enhancement has improved the diagnostic performance.

Image enhancement may also be performed as a preprocessing step in an automated image analysis operation. The enhancement in this case will often simplify an image considerably with the result that it may not appear particularly interesting to the human observer, but would make further automated image enhancement or analysis processes viable.

Grayscale image enhancement is covered in Chapter 1 and enhancement of the spatial properties in Chapter 3.

### Activity: Windowing in ImageJ

Windowing is an option available on many scanner consoles. A selected range of gray values is remapped to display over the full range from 0 to 255.

Start ImageJ and load the image spine.bmp from the CD.

Select Image | Adjust | Window/Level ....

Change the Level to 120 and the Window to 160.

Watch the effect on the image when the Reset button in the W&L window is clicked to return the values to 128 and 255.

See the effect of the Auto button, then Reset again.

Adjust the Window slider, moving it to the right hand end of the scale and watching the effect on the image. This kind of extreme contrast enhancement is the kind that may be of value in automated image analysis.

Close the W&L window, close the image and exit from ImageJ.

### 1.6.2   Image Restoration

Image restoration is required if the image acquisition method causes geometric distortion, blurring or inhomogeneity in the image. Geometric distortion means that square objects are imaged as nonsquare shapes, either pulled in (pincushion distortion) or puffed out (barrel distortion). Images from MRI, for example, may suffer geometrical distortion at the extremes of the field of view. This distortion may not be noticeable to the eye but is important if, for example, MRI and CT data are to be combined. Image restoration methods are covered in Chapter 7.

### 1.6.3   Image Analysis

The final results from image analysis operations are usually not themselves images, but are numerical measurements made from the image data. Examples of image analysis operations include measurement of length, area or volume, object classification, image segmentation, and image arithmetic. Basic measurement methods were covered in Chapter 1, classification and segmentation are covered in Chapter 2, and logical and arithmetical operations are in Chapter 5.

### 1.6.4  Image Compression

Image data files are often very large in terms of the number of bytes they have. For example, a CT examination of 100 images might occupy up to 40 Mbytes (megabytes). The large size not only has implications for the amount of disk space required, but is important if images are to be transferred elsewhere, for example, for teleradiology. The larger the file, the longer it takes. Two types of compression algorithm are available for reducing image file size: lossy and lossless. Lossless compression is the only one that should be used in medical image processing, because exactly the same image data are available on decompression, whereas the data are irreversibly changed when using lossy techniques. Image data compression methods are covered in Chapter 6, together with the related topic of medical image data formats. Data formats describe standardized ways of storing image and patient data.

### 1.6.5  Image Synthesis

Image synthesis includes activities such as image registration, multidimensional visualization and 3-D rendering. Image registration is a very important technique that can be used quantitatively to combine information from several different sources. There are many reasons for performing such an operation, including the assessment of disease progression or growth using temporal series and to combine structural and functional information from different modalities. A 3-D rendering is the visualization of a 3-D view, often incorporating cues such as depth and shading to make an image look more three dimensional. Image registration is covered in Chapter 8, and visualization in Chapter 9.

## 1.7  Good Practice Considerations

Subjective windowing of an image is a valuable tool for personal viewing of an image, but it is important to keep subjective image enhancement operations separate from any protocol that leads to quantitative results.

In quantitative image processing, good practice demands that the operator is always able to justify the choice of any operation and ensures that parameters are chosen in a way that makes them applicable to every processed image.

Keep the original image safely stored and always work on copies. This is important to remember even when just trying things out before a formal analysis. In this scenario, before a full analysis has even begun, one can easily make a silly mistake and overwrite important data.

## 1.8 Chapter Summary

In this chapter the basic principles and definitions of image processing were described. Spatial and grayscale properties were addressed separately. The ImageJ image processing program was introduced and a number of activities using the program were included. Brief descriptions of the five classes of image processing were given and used to provide an overview of the content of the book.

## 1.9 Feedback on the Self-Assessment Questions

### Self-assessment question 1.01

The image of the shoulder joint is a magnetic resonance (MR) image. Loss of image information on the side of the image away from the body surface is characteristic of an image acquired using a surface coil.

### Self-assessment question 1.02

There would be $2^5$ shades of gray, (i.e., $2\times2\times2\times2\times2 = 32$).

### Self-assessment question 1.03

The word "pixel" comes from picture element.

### Self-assessment question 1.04

Figure 1.8b has the higher spatial frequency.

### Self-assessment question 1.05

- There are two gray values in a binary image, usually displayed as black and white.
- If n is the number of bits, then for both signed and unsigned data types, $2^n$ values can be represented. Unsigned data types use values between 0 and $(2^n-1)$, so an unsigned 8-bit scale runs from 0 to 255. A signed 8-bit scale usually runs from –127 to 128, so it cannot display

such large positive values as the unsigned 8-bit scale. For a signed data type, 0 is usually shown as a mid-gray, between black for negative pixel values and white for positive.

- Integer numbers are most easily stored in a computer in either 1-byte or 2-byte format, so it is most common for images to be 8-bit or 16-bit images.

### Self-assessment question 1.06

The binary image appears in Figure 1.9b. A binary image has only two gray values. Note that, as here, they need not be 0 and 1, or 0 and 255, or black and white, though they often are. The other images have 4, 256 and 8 levels, respectively.

### Self-assessment question 1.07

The dynamic range of a gray scale is the range of pixel values used (e.g., 0 to 255 for 8-bit scale).

### Self-assessment question 1.08

The images in Figure 1.13a and d have the same spatial resolution and this spatial resolution is the highest shown, (c) is next and (b) has the lowest spatial resolution.

### Self-assessment question 1.09

The spine image, Figure 1.16a, has the lower image contrast. It is hard to distinguish details in the image.

### Self-assessment question 1.10

Figure 1.23a: A predominantly dark image.
Figure 1.23b: A predominantly light image.
Figure 1.23c: An image with a small dynamic range.
Figure 1.23d: An image with a wide dynamic range.
Figure 1.23e: A well-balanced image with a large dynamic range.
Figure 1.23f: A high-contrast image with a large dynamic range.
The images corresponding to each histogram are shown in Figure 1.40.

### Self-assessment question 1.11

The dark background pixels in medical images can spoil the results of histogram equalization if they are not accounted for. The background pixels will be treated the same as any other pixel value, so the large number of pixels with low gray values can be spread across a larger number of gray values, and the result is a brighter and noisy background, with the pixels that are

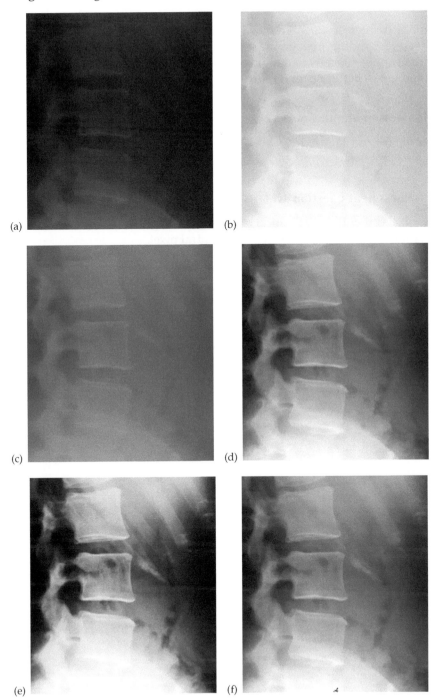

**FIGURE 1.40**
Images corresponding with each histogram in Figure 1.23. (a) A predominantly dark image, (b) a predominantly light image, (c) an image with a small dynamic range, (d) an image with a wide dynamic range, (e) a well-balanced image with a large dynamic range, (f) a high-contrast image with a large dynamic range.

actually of interest squeezed into a small range of high values (Figure 1.36a). The corresponding histogram is shown in Figure 1.36b. Histogram equalization, applied locally to a region of interest that excludes the background, should give a better result (Figure 1.37).

### Self-assessment question 1.12

Figure 1.36 and Figure 1.37 show the expected results without and with a region of interest.

### Self-assessment question 1.13

The pixels other than the central one are the 8-neighbors of the central pixel. They are said to be 8-connected to the central pixel.

## Reference

1. The Chapel Hill Volume Rendering Test Data Set, Volume I, Department of Computer Science, University of North Carolina. Available at http://education.siggraph.org/resources/cgsource/instructional-materials/volume-visualization-data-sets

# 2

## Segmentation and Classification

## 2.1  Introduction

Segmentation and classification are image processing operations that fall into the image analysis category. Both are concerned with dividing an image into regions or components that have a meaning associated with the particular application. In medical image processing, for example, segmentation might be performed to separate a lesion from the rest of an image so that measurements can be made. In classification, more types of region are included, so that every pixel or voxel is defined as belonging to a class.

### 2.1.1  Learning Objectives

When you have completed this chapter, you should be able to

- Define segmentation in image analysis
- Give examples of manual, automatic and semiautomatic methods of segmentation
- Discuss factors affecting the reproducibility of segmentation methods
- Define classification in image analysis
- Give examples of methods of classification
- Use ImageJ to perform image segmentation

### 2.1.2  Example of an Image Used in this Chapter

*Self-assessment question 2.01*

Figure 2.1 shows a CT section through the abdomen where a contrast agent has been used to emphasize the lumen of the aorta. Identify this area on the image.

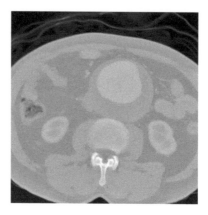

**FIGURE 2.1**
Image for self-assessment question 2.01 showing an x-ray CT section through the abdomen.

## 2.2   Segmentation

Segmentation is an image analysis technique. The term *segmentation* is usually used to describe the process of selecting a specific object or objects from an image. It is usually followed by some further operation, for example to determine the volume of the object or the generation of a visualization.

Segmentation is not identical to the process of classification, which is covered later in this chapter, although the processes can sometimes be closely linked. The classification process involves assigning each pixel in the image to a class, depending to which object it belongs.

### 2.2.1   Approaches to Segmentation

There are three approaches to segmentation

- Manual
- Automatic
- Semiautomatic

Manual methods are very commonly used. For example, an expert radiologist will use an image processing program to trace the boundary of an organ seen in CT slices. All decisions about what should, or should not, be included in the segmented region are made by the human observer. This means that, given a suitable tool, work on segmentation can start as soon as the images are acquired. This is in contrast to the automated and semiautomated approaches that are covered later in this chapter, where it is necessary first to acquire sufficient data to develop and test the techniques. As a result, manual methods are often used in pilot studies.

However, manual methods have many drawbacks:

- they are very time-consuming
- they are subject to human error
- they are subjective (i.e., have poor interobserver reproducibility)
- they have poor intraobserver reproducibility

These drawbacks are the motivations for automatic segmentation. In automated techniques, computer-based methods are devised that follow strict rules and exclude human intervention. Automation should overcome the reproducibility limitations of manual methods (at least on a single image), but they still tend to be prone to errors, which arise because of the variations between images that occur in a study. It is quite possible for an automated technique to work perfectly on a set of example images, known as the training set, but to fail on further examples because they violate an assumption that was true for all images in the training set. This might happen if a structure that was well separated in all the training images was much closer to the object of interest in the example used, or if images had different contrast or noise characteristics because they came from a more diseased individual than any in the training set.

Semiautomatic segmentation is the compromise between the two approaches. The human observer usually starts the segmentation, and has the opportunity to make corrections, but as much of the work as possible is automated. These techniques may also be known as assisted manual segmentation.

*Self-assessment question 2.02*

It is common for image analysis to be used in studies that test the efficacy of a treatment or accuracy of a diagnosis. Multi-center studies may be used to increase the number of subjects in the study. What is the principal reason for automated or semi-automated methods being preferred over manual methods of segmentation for multi-center studies?

## 2.2.2 Manual Segmentation Methods

Manual segmentation involves an expert observer outlining the object of interest in the image. Nowadays this is done on a computer monitor, rather than on paper or film. The computer simply provides tools to help with drawing that outline; with the manual methods there is no attempt by the computer to influence where the line is placed. Software may offer a choice between drawing an *open* or *closed contour*. In the latter case, when drawing around an object there is no need to ensure that the ends of the tracing meet, as they will automatically be joined. This is the most common mode to use, with open contours applied if closed ones fail to work as required. In ImageJ, the open and closed options are in the Tools menu as indicated by the arrows in Figure 2.2. An example of a closed contour drawn in ImageJ is shown in Figure 2.3.

**FIGURE 2.2**

The Tools menu in ImageJ, with the open and closed freehand selection options indicated by the white arrows. In recent versions of ImageJ, when the icon is labeled with a small arrow, it is necessary to use the right mouse button to see the alternatives available.

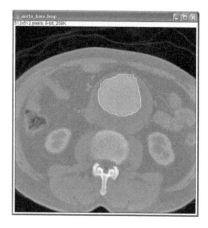

**FIGURE 2.3**

An example of a closed contour drawn using the freehand selections tool in ImageJ.

*Self-assessment question 2.03*

Which of the two arrowed symbols in Figure 2.2 represents the closed contour option?

In software packages, the ability to click on a section of the outline and pull it to the right place, with the rest of the outline unaffected, is usually implemented using a mathematical process called *spline* fitting. The outline is converted to a series of *control points* around the shape, which are connected by a mathematical function (the spline) that matches the drawn outline. The software allows these control points to be moved, and the curve smoothly adapts to fit the new position of the control point (Figure 2.4).

As indicated earlier, manual segmentation is notorious for its poor reproducibility. For good reproducibility, the operators need to be given explicit instructions about their task. It should be explicitly stated whether they should they be drawing on the outermost pixels of the object, or around the outside of the object on pixels that represent nonobject material. Clear instructions should be provided regarding the definition of the boundaries of anatomical features, and a standard protocol set out regarding viewing conditions and image enhancement.

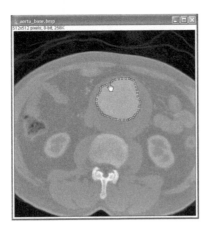

**FIGURE 2.4**
An example of a closed contour with a spline fitted in ImageJ.

**FIGURE 2.5**
The ImageJ freehand selections icon in the ImageJ Tools menu.

## Activity: Manual Segmentation in ImageJ

Start ImageJ and load the image prosthesis.bmp from the CD.

Select the freehand selections icon in the ImageJ Tools menu (Figure 2.5).

The prosthesis is the bright object in the image. Click and hold down the mouse button somewhere on the edge of the prosthesis. Continue to hold down the mouse button and draw around the edge of the object. When the button is released, the two ends of the line will be joined. Drawing in this way is not easy, and several attempts may be required. To draw a new outline, click once anywhere in the image outside the outline, then start to draw again.

When a satisfactory outline has been drawn, select Edit | Clear Outside. This will set parts of the image outside the defined region of interest to zero.

If the background area does not go black, select Edit | Undo to return the image to its previous condition. Then select Edit | Options | Colors, and ensure that the settings are Foreground: White and Background: Black. Repeat the Clear Outside operation.

Select Edit | Undo, then see what happens when Edit | Clear is used instead of Clear Outside.

Note that in both cases the original pixel values are retained in areas not set to zero. The Edit | Fill option sets selected pixels to the Foreground color.

*Self-assessment question 2.04*

All but one of the factors in this list will affect the reproducibility of manual segmentation. Identify the inapplicable factor.

> Degree of experience of the operator
> Color of chair upholstery
> Level of room lighting
> Brightness and contrast settings on monitor
> Operator fatigue

*Self-assessment question 2.05*

The three images in Figure 2.6 should suggest to you two factors that affect the reproducibility of manual segmentation. What are these two factors?

### 2.2.3    Automatic and Semiautomatic Segmentation

#### 2.2.3.1    Thresholding

Thresholding on gray level is one of the simplest ways of introducing objectivity into the segmentation process. Thresholding involves choosing a range of gray levels that represents the object to be segmented. Only those pixels in the defined range are retained, and all the others are set to a fixed value, which is often zero. Sometimes, the selected pixels are set to have a specified value such as 0, 1 or 255. The image in Figure 2.7a has been thresholded so that all pixels with a value between 142 and 255 have been displayed in white.

For quantitative image analysis, it is necessary to devise an objective and justifiable method for choosing the threshold. Examples of this approach include using a percentage of the maximum pixel value in the image, which will compensate for differences in brightness between images from different individuals, or selecting a value based on analysis of the image histogram. For the image in Figure 2.7b, the chosen threshold was based on analysis of the histogram of the elliptical region of interest.

*Self-assessment question 2.06*

Which of the following terms describes an image that contains pixels which have only two possible values? (More than one of the options is correct)

> 8-bit
> Binary
> 1-bit
> 2-bit
> Doppler

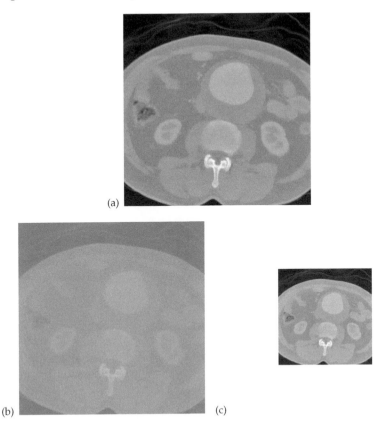

**FIGURE 2.6**
(a)–(c) Images for self-assessment question 2.05

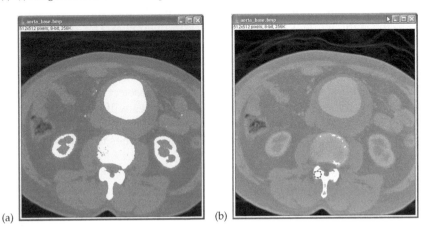

**FIGURE 2.7**
Gray level thresholding. (a) Pixels with values 142–255 are displayed in white, (b) the mean and standard deviation of the gray values within an elliptical region were found. The threshold level was set at two standard deviations below the mean, and calculated from the measurements. Pixels with values greater than the threshold are displayed in white.

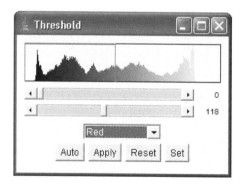

**FIGURE 2.8**
The ImageJ Threshold dialog.

### Activity: Thresholding in ImageJ

Start ImageJ and select Edit | Options | Colors. Ensure that the settings are Foreground: White and Background: Black.

Load the image prosthesis.bmp from the CD.

Move the cursor across the image and view the gray values that are reported in the status bar at the bottom left of the ImageJ menu. The values are higher in the prosthesis than in the surrounding area.

Select Image | Adjust | Threshold. The ImageJ Threshold window (Figure 2.8) will appear. Move the slider in the lower of the two bars (which represents the maximum of the range to be selected) as far to the right as possible. The number displayed should be 255. Slowly move the other slider (in the upper bar) to the right, while watching the effect on the image. The aim is to select (color red) all of the prosthesis but include little of the rest of the image (Figure 2.9a). The range 200–255 is a good choice.

Click on the Apply button. In the Apply LUT dialog which appears, include ticks beside the first two statements, and deselect "Black foreground, White background." This combination of responses will result in an image where the prosthesis and other thresholded areas are white and the rest of the image is set to zero (black) (Figure 2.9b). Sometimes the white area will be shown in red.

### 2.2.3.2   Region Growing

Region growing is an interactive way of defining a region of interest. The operator defines a range of gray levels that represents the region, and chooses a pixel that is known to be part of the region. The chosen pixel is known as the seed point. All the pixels that have a gray level within the range and are connected to the seed point are selected as being in the region. In a variation on the method, an outline is drawn automatically around the set of pixels, so

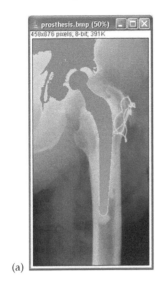 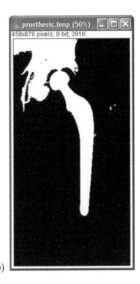

**FIGURE 2.9**
Thresholding the image prosthesis.bmp. (a) The threshold gray level range is adjusted and the display changes to show the selected areas in gray (usually red when using ImageJ). (b) The result of applying the chosen range of gray levels for thresholding.

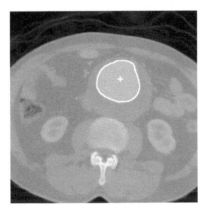

**FIGURE 2.10**
Region growing. The cross indicates the seed point which is placed inside the required region. The outline drawn around the set of qualifying pixels in the range 144–255 is shown by the white line.

that the selected region is identified by a line rather than by a set of pixels. In Figure 2.10, the cross indicates the seed point, and the white line enclosing the seed point and cross is the outline drawn around the set of qualifying pixels in the range 144–255.

Segmentation methods that rely on the gray levels within the image will be adversely affected by *gray level inhomogeneity* and the *partial volume effect*. Inhomogeneity (Chapter 7) means that the same tissue may not be represented by the same gray value everywhere in the image. The partial volume effect arises from the possibility that a medical image pixel or voxel is large enough to contain more than one type of material. Low spatial resolution modalities are most affected. The consequence is that the distinction in gray levels between materials is blurred and boundaries are not well defined.

### 2.2.3.3 Active Contours and Snakes

The mathematics that underlies the active contours and snakes methods is quite complicated, but tools are now available that allow them to be used successfully by the naive user. Both methods involve using an outline, or contour, to represent the border of the region of interest. The outline starts out very roughly positioned in the image, and then calculations are performed that drive the contour towards points in the image where the image satisfies predefined conditions. The conditions are based on known properties of the boundary of the desired region of interest, perhaps in terms of edge sharpness or a contrast ratio across an edge. Additional conditions may also be applied that restrict the contour to take up only realistic shapes. This variant of the active contour technique is called a *deformable model* or *active shape model*. Models such as these need to be specially developed for a particular segmentation task; a number of images will be used to *train* the model and thus define the range of realistic shapes.

Figure 2.11 shows a snake being used on an x-ray image of the hand. The segmentation task is to find the edges of the largest bone in the image. Figure 2.11a shows the initial ellipse-shaped contour, and Figure 2.11b is the result after the snake program has run. The contour has successfully moved to the edges of the bone, but, because there are no shape constraints, at the ends of the bone, it has been attracted to an edge that is part of a different bone. This characteristic of snakes means that the initial contour (the ellipse

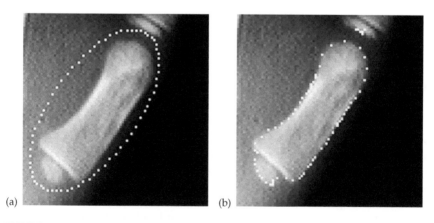

(a)                                                      (b)

**FIGURE 2.11**
Snake used for segmentation of bones in an x-ray image of the hand. (a) Initial ellipse-shaped contour, (b) snake fitted to the bone edges, but with some errors. (Courtesy of Nick Efford. With permission).

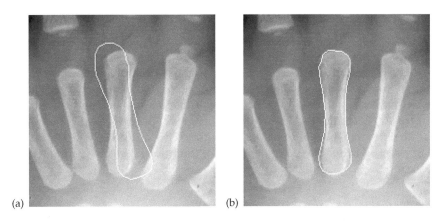

**FIGURE 2.12**

Deformable model used for segmentation of bones in an x-ray image of the hand. (a) Initial bone-shaped contour, (b) contour fitted to the bone edges. (Courtesy of Nick Efford. With permission.)

on the left in this case) needs to be placed very close to the correct edge to start with. The deformable, or active shape, model overcomes this drawback by including knowledge of the shape that is expected. In the example in Figure 2.12, the initial contour (Figure 2.12a) is clearly bone shaped. In the resulting image after the program has been run (Figure 2.12b), the shape has fitted itself neatly around the edges of the bone.

Computational segmentation methods are areas of active research, with new techniques being developed and made available all the time. A degree of understanding of any method is necessary to ensure that it is appropriate to the data and for the selection of sensible values for the parameters.

### 2.2.3.4 Mathematical Morphology

Morphological image processing techniques are often used to refine the binary image that is the result of segmentation. The techniques are valuable for removing noise or creating an outline. Morphological operations are covered in Chapter 5.

*Self-assessment question 2.07*

The following terms are all important for analytical methods: accuracy, precision, interobserver reproducibility and intraobserver reproducibility. Is accuracy or precision synonymous with reproducibility?

## 2.3 Classification

Classification is an image analysis technique in which every pixel or voxel in an image is assigned to a particular *class*. This process is sometimes called *labeling.* In medical images the classes will normally represent tissue types, such as gray matter or white matter in the brain. There may be a background class. Classification often incorporates spatial information (sometimes called context), so pixels in a class not only share a gray level property, but are also close to each other in the image. Once classification has been performed, one could use the results for segmentation, for example, by choosing to set to zero all pixels other than those in the required class. Alternatively, one might retain the full set of classifications, to use the information in an anatomical atlas or for a 3-D visualization.

### 2.3.1 Multispectral Classification

A set of classification methods, which may be called *multispectral classification* or *cluster analysis*, relies on there being more than one numerical property or *feature* associated with each voxel. In other words, there must be two or more images representing the same object. A good example is from magnetic resonance imaging, where different pulse sequences may be used to generate images of exactly the same object but with different appearances. The more of these properties or features that voxels have in common, the more likely it is that the voxels belong to the same class. The inclusion of information related to more than one property can help resolve confusion that arises if pixels from different classes share the same value of one image property. This is illustrated in Figure 2.13, Figure 2.14 and Figure 2.15. Figure 2.13

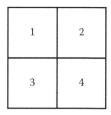

**FIGURE 2.13**
An object comprising four different classes of material (1, 2, 3, and 4).

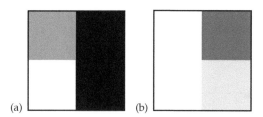

**FIGURE 2.14**
(a) Image A of the object in Figure 2.13 acquired using first imaging technique. (b) Image B of the same object acquired using a second imaging technique.

**TABLE 2.1**

Pixel Values for the Four Classes
in Images A and B (Figure 2.14)

|         | Image A | Image B |
|---------|---------|---------|
| Class 1 | 128     | 255     |
| Class 2 | 0       | 95      |
| Class 3 | 255     | 255     |
| Class 4 | 0       | 191     |

shows an object made up of four classes of material: classes 1, 2, 3, and 4. The two images in Figure 2.14 represent two different images acquired of this object. In Image A, classes 2 and 4 have the same gray value (gray value is the *feature* in this case), and in Image B, classes 1 and 3 cannot be differentiated. Neither image used alone will differentiate the four classes. However, the four classes can be differentiated by using information from both the images. Table 2.1 shows the pixel values for the four classes in images A and B, and it can be seen that classes 2 and 4, which are indistinguishable in Image A, will quickly be resolved by using data from Image B, too.

Another way of visualizing these data is graphically, using one axis to represent pixel values in image A and the other for image B (Figure 2.15). It can be seen that by using information from both images, the overlapping classes are pulled apart. This method of plotting feature values is called a *scattergram*, or sometimes *feature space*.

A feature space can have more than two axes, or dimensions. In the previous example, there would be an additional dimension if there was a fifth class of material in the object, which had the same values as one of the other classes in both images A and B, but could be identified by information from a third

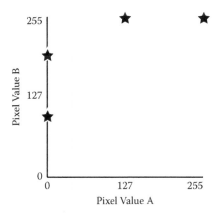

**FIGURE 2.15**
Scattergram, or feature space, plotted for the pixel values seen in the images A and B. By using information from both images, all four classes are separated.

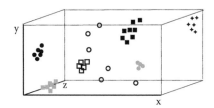

**FIGURE 2.16**
Sketch of a scattergram showing how feature values group together in clusters in feature space.

type of image. The scattergram would have three axes. Multidimensional feature spaces are common, and can be analyzed, but because it is hard to represent more than three dimensions in a graph, they are not generally plotted.

The principle behind image classification techniques is the use of more than one type of image of an object to differentiate better between classes. In real situations, the gray levels will not be exactly the same for all pixels of the same class; instead, there will be some spread around a mean value. So when the scattergram is plotted, there will be *clusters* of points with approximately the same pixel values (Figure 2.16). Clustering algorithms are used to assign each pixel in an image to a particular cluster. There are many different clustering algorithms available. Whichever clustering algorithm is used, for the classification to work, the following assumptions are made:

- The images are in perfect spatial registration so that pixels at the same position in each image represent the same physical location (Chapter 8).
- There is gray level homogeneity across the image. If a particular material is represented by a particular gray value in one place in the image, then it is represented by the same gray value everywhere else in the image, too (Chapter 7).

*Self-assessment question 2.08*
Which of the images in Figure 2.17 are affected by gray level inhomogeneity?

## 2.3.2   Supervised vs. Unsupervised Classification Methods

*Supervised* classifiers are run after the user has manually identified samples of parts of the image known to belong in the different classes. The classifier then works by assigning other pixels to the classes by determining how similar each one is to the defined samples. *Unsupervised* classifiers form clusters of values without any prior knowledge about where in feature space they will be. The user usually defines only how many classes there should be and perhaps limits on their extent. This sort of process is described as "data-driven" because the results are dependent on the data.

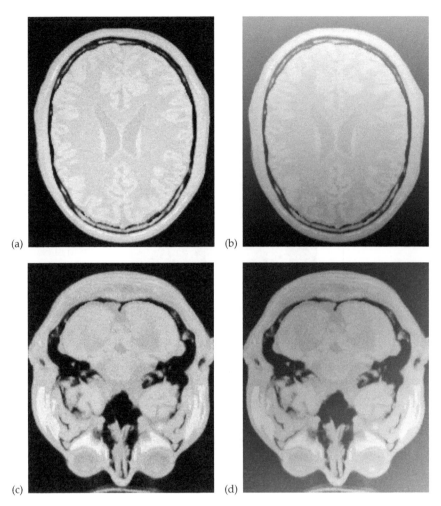

**FIGURE 2.17**
(a)–(d) Images for self-assessment question 2.08. (Image data from the BrainWeb Simulated Brain Database.[1,2] With permission.)

### 2.3.3 Medical Imaging Applications for Classification

The medical image data to which these classification methods may be applied include:

- MRI data, for example, different pulse sequences are used to give T1-weighted, T2-weighted and proton density weighted images. Applications using MRI have included brain tumor measurement, MS lesion identification and CSF volume measurement.
- Two different terahertz pulsed imaging images of the same object are shown in Figure 2.18 together with a classified result.

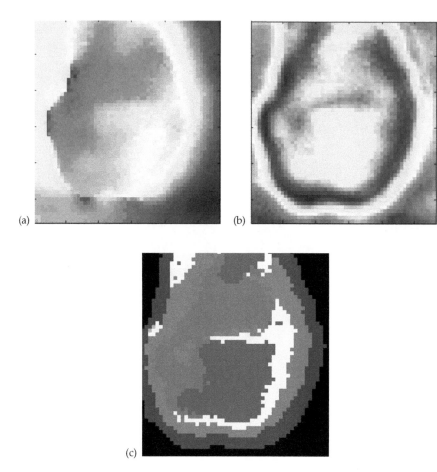

**FIGURE 2.18**
Terahertz pulsed imaging data and multispectral classification. (a)–(b) Two different images calculated from analysis of transmitted pulses, (c) classification using an unsupervised algorithm. (Adapted from Berry, E. et. al., Multispectral classification techniques for terahertz pulsed imaging: an example in histopathology, *Med. Eng. Phys.*, 26, 423, © 2004. With permission from IPEM.)

- X-ray computed tomography (CT) images may be combined with positron emission tomography (PET) or MRI data, a process that has been facilitated by improved methods for image registration (Chapter 8).

- Confocal microscope images have three color channels (red, green and blue channels) that may be combined using clustering methods.

### 2.3.4  An Example of Supervised Classification

In this example the MR images are of a simulated brain with simulated MS lesions. The two images chosen are T2-weighted and T1-weighted slices,

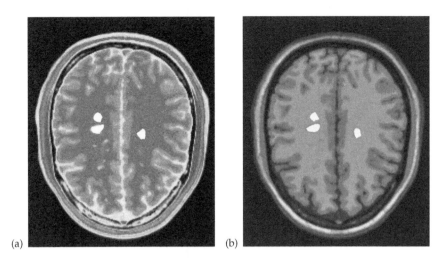

(a)  (b)

**FIGURE 2.19**
MR images to be used for supervised classification. Areas known to represent lesions have been highlighted in white by the operator. (a) T2-weighted, (b) T1-weighted. (Image data from the BrainWeb Simulated Brain Database [1,2]. With permission.)

shown in Figure 2.19a and b. Several areas known to contain lesions have been highlighted in white by the operator as the supervised part of the classification process. A simplified scattergram is shown in Figure 2.20, with the area occupied by the pixels that have been selected in the two images shaded in gray and the area occupied by all the others pixels outlined. The selected pixels appear to occupy a single region separated from the areas occupied by other pixels, which is promising for multispectral analysis. Figure 2.21 shows the scattergram after the *k-nearest neighbors* clustering algorithm has been applied to identify pixels that should be members of the lesion class that was identified by the operator. The shaded region of the scattergram is associated with the lesion class. In this case a maximum distance parameter of the algorithm was set to the value 10, and the region would be smaller for a smaller value of the maximum distance. In practice it is necessary to use training data to determine the values to use. Figure 2.22a shows the result of identifying the pixels in the image pair that are associated with the lesion class. As there is only one class, this is the equivalent of a segmentation step. Figure 2.22b is the binary "ground truth" image showing the simulated lesions that are known to be present. To compare the calculated result with the ground truth, one image is subtracted from the other. Image subtraction is covered in Chapter 5. In Figure 2.22c, gray is the value 0 and represents pixels that appear in both images. White pixels are those that are in the first image and not the second, and black pixels are in the second image but not the first. In this case the second image was the ground truth image.

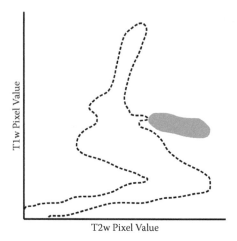

**FIGURE 2.20**
Simplified scattergram associated with the images in Figure 2.19. Gray shading indicates the areas occupied by the pixels that were highlighted in white by the operator. The dashed line encloses the area occupied by the other pixels.

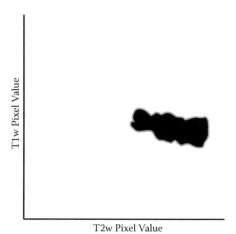

**FIGURE 2.21**
Simplified scattergram after running the *k-nearest neighbors* clustering algorithm. All pixels in the black area are members of the same class as those initially highlighted by the operator.

## 2.4   Good Practice Considerations

Semiautomated methods for segmentation will be prone to subjectivity and have poor reproducibility if insufficient care is taken when selecting values for parameters. The method and justification for all selections must be recorded and reported.

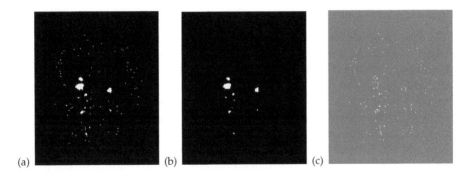

**FIGURE 2.22**
(a) Classified image showing the pixel locations corresponding with the black area of Figure 2.21. (b) Ground truth image showing the known locations of the simulated lesions. (c) Result of subtracting the ground truth image from the classified image. White and black pixels represent pixels that have been wrongly classified. (Image data from the BrainWeb Simulated Brain Database [1,2]. With permission.)

It is important to ensure that a clear protocol, both for acquiring the original images and for performing the analysis, is devised and adhered to.

If a complex built-in algorithm is used, such as a clustering algorithm, ensure that the relevance of the selectable parameters is understood so that the choice of both algorithm and parameter values may be fully justified.

## 2.5 Chapter Summary

This chapter was concerned with two topics, segmentation and classification, from the image analysis class of image processing. The concepts of manual, automated and semiautomated segmentation were explained. ImageJ activities covered simple manual and semiautomatic approaches. The principles underlying classification by clustering were outlined to provide readers with an understanding of the key terms likely to be encountered in this field. Issues associated in evaluating the quality of segmentation and classification results are addressed in Chapter 10.

## 2.6 Feedback on the Self-Assessment Questions

### Self-assessment question 2.01

The lumen of the aorta is the large, bright, almost circular region in the upper half of the image. It would not be so well differentiated if the contrast agent had not been used.

### Self-assessment question 2.02

The principal reason that automated or semiautomated methods are preferred over manual methods of segmentation for multicenter studies is their better reproducibility. In multicenter studies it is very important to ensure that the data from all the centers are treated in the same way, but interobserver variability can be a major problem. It is also important to minimize other differences between the centers by ensuring that the same imaging protocol is used in each, and, where possible, the same equipment. It is certainly true that manual methods are time consuming, but observers are often willing to devote long periods to a study, particularly if it will be the first of its kind and lead to a publication that is widely cited. One reason that those designing a clinical study may not wish to use nonmanual methods is that these methods take time to be developed and evaluated. Most researchers are keen to publish their results as soon as possible. As a result, it is common for work to proceed in parallel. Manual image analysis methods will be used for the measurements for initial publications, and automated techniques will be developed at the same time. Testing of the automated methods may use the manual results as the gold standard reference.

### Self-assessment question 2.03

The heart-shaped icon on the left represents the closed contour. Those familiar with using different kinds of software were probably able to guess from the icon itself. If not, it takes just a few moments to experiment.

### Self-assessment question 2.04

The most likely inapplicable factor affecting the reproducibility of manual segmentation is color of chair upholstery. However, it was very hard to think of an inapplicable factor to list in this question because human operators can be affected by many things. The large range of potentially confounding factors contributes to poor reproducibility for manual methods.

### Self-assessment question 2.05

The two factors, illustrated in this question, which affect the reproducibility of manual segmentation are display contrast and displayed size. In the image that has reduced contrast compared with the others, it is much harder to see the boundaries of the objects. Similarly, the half-size image would be harder to draw on. It is possible to magnify the image display to see it better, but if the edges of pixels become apparent as a result of magnification, then the magnification is too large.

### Self-assessment question 2.06

The two correct terms to describe an image containing pixels that have only two possible values are Binary and 1-bit. A 1-bit image has $2^1$ (= 2) possible gray levels, 0 and 1. A binary image contains just two gray levels; often these are 0 and 1, but they do not need to be. An 8-bit image containing pixels with values of only 0 and 255 is also a binary image. A 2-bit image has $2^2$ (= 4) possible gray levels, and an 8-bit image has $2^8$ (= 256) possible gray levels. Doppler is not used in this particular imaging context.

### Self-assessment question 2.07

Precision is synonymous with reproducibility. Precision indicates the degree to which a method gives the same answer on repeated measurements. Note that the answer concerned need not be the correct one; the requirement is for repeated measurements to be close together. In contrast, accuracy is the term that defines how close the result is to the known correct answer, or ground truth. See Chapter 10 for a more detailed consideration of the terms.

### Self-assessment question 2.08

Images (b) and (d) in Figure 2.17 are affected by gray level inhomogeneity. In Figure 2.17b, note how gray and white matter at the top of the image is brighter than the same tissues at the bottom. In Figure 2.17d, the bottom right of the image is brighter than the top left, leading to differences in the appearance of similar tissues, which are arrowed in Figure 2.23. Images affected by gray level inhomogeneity would not work well for clustering because the same tissue is not always represented by the same gray level (feature value). This might lead to clusters containing two different tissues.

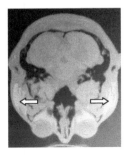

**FIGURE 2.23**
Image affected by gray level inhomogeneity, similar tissues (arrowed) have different gray levels. Images affected in this way would be unsuitable for clustering analysis unless the inhomogeneity was first corrected (Chapter 7). (Image data from the BrainWeb Simulated Brain Database [1,2]. With permission.)

## References

1. The BrainWeb Simulated Brain Database. www.bic.mni.mcgill.ca/brainweb/
2. Collins, D.L. et. al., Design and construction of a realistic digital brain phantom, *IEEE Trans. Med. Imag.*, 17, 463, 1998.

# 3

## Spatial Domain Filtering

### 3.1 Introduction

The operations in this chapter are sometimes termed neighborhood operations because the value of a pixel is changed according to calculations involving the value of the pixel itself and the values of nearby pixels. The operations fall into the image enhancement category of image processing, and are also important as preliminary steps, or preprocessing operations, in image analysis.

#### 3.1.1 Learning Objectives

When you have completed this chapter, you should be able to

- Define rank filtering
- Define convolution filtering
- Give examples of simple rank and convolution kernels
- Discuss the effect of the size and shape of the neighborhood used for filtering
- Use ImageJ to perform a range of spatial filtering operations
- Recognize the need to choose operations and parameter values in a justifiable way

#### 3.1.2 Example of an Image Used in This Chapter

*Self-assessment question 3.01*

Figure 3.1 is an x-ray computed tomography image of the head. Give two ways in which a transverse T2-weighted magnetic resonance image of the same slice would differ.

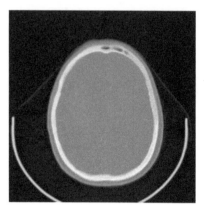

**FIGURE 3.1**
Image for self-assessment question 3.01 showing an x-ray computed tomography image of the head.

## 3.2 Spatial Filtering Operations

The neighborhood spatial filtering operations fall into two categories: *rank* filtering and *convolution* filtering. A third *hybrid* category involves a combination of operations.

### 3.2.1 Rank Filtering

In rank, or ordered, filtering the gray values of pixels within a defined neighborhood around the pixel of interest are simply arranged in a list in ascending order. The new value for the pixel of interest is the value at the required rank position in this list; for example, the largest value would be chosen if maximum filtering were being performed. The defined neighborhood may be square, as illustrated in Figure 3.2, but other shapes may be chosen too.

Four commonly available rank filters are the median, maximum, minimum and range filters. Median filtering is valuable when an image is corrupted by noise characterized by noise pixels having a very much higher or lower value than the rest of the image. This is sometimes termed impulse noise, where noise pixels have values from the extremes of the range, e.g., 0 or 255 for an 8-bit image. A median filter will replace any value that is extreme for its neighborhood by the median value in the ranked list, thus removing this kind of noise pixel without excessive blurring, as illustrated in Figure 3.3.

The maximum and minimum filters are easy to understand: each pixel in the image is replaced by the largest, or smallest, value in its neighborhood. The effect on dark and bright areas is illustrated in Figure 3.4. The original image is the same noisy image of Figure 3.3a. The minimum filter enlarges the size of dark areas, while the maximum filter enlarges light areas. The images are also slightly blurred.

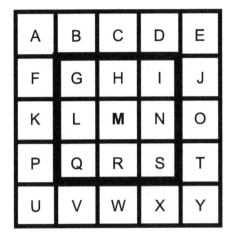

**FIGURE 3.2**
A square 3×3 neighborhood for rank filtering. The result calculated from the pixel values within the square will replace the value M.

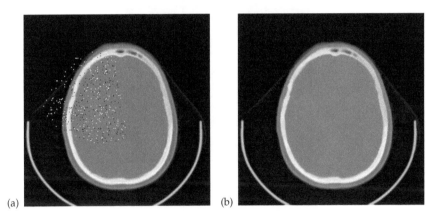

(a)                                             (b)

**FIGURE 3.3**
(a) Original image affected by impulse noise, (b) result of applying a 3×3 square median filter.

The range filter acts as an edge detector. Each pixel is replaced by range of the gray values in the neighborhood, that is, the difference between maximum and minimum values. Near the edge of an object, the range in the neighborhood will be larger than inside an object, hence the edge detection characteristic of the filter. An example is shown in Figure 3.5.

*Self-assessment question 3.02*

The pixel values for a 5×5 portion of an image are shown in Figure 3.6. Four pixels have been marked with the letters a–d. If a square 3×3 neighborhood is used, what is the result of applying the maximum and the minimum filter to each marked pixel?

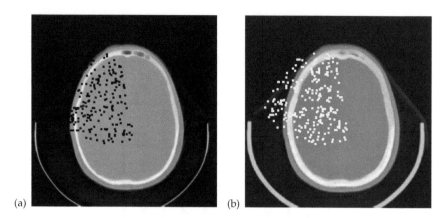

(a)                                              (b)

**FIGURE 3.4**
Result of applying rank filters to the original image of Figure 3.3a. (a) 3×3 square minimum filter, (b) 3×3 square maximum filter.

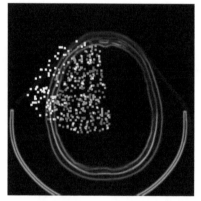

**FIGURE 3.5**
Result of applying a 3×3 square range filter to the original image of Figure 3.3a.

| 98 | 112 | 134 | 154 | 169 |
|---|---|---|---|---|
| 89 | 89 (a) | 169 | 169 (b) | 169 |
| 89 | 89 | 89 (c) | 169 | 169 |
| 89 | 89 (d) | 89 | 89 | 89 |
| 89 | 89 | 89 | 89 | 89 |

**FIGURE 3.6**
Image for self-assessment question 3.02.

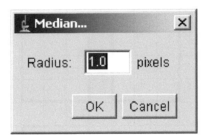

**FIGURE 3.7**
The ImageJ Median Filter dialog.

## Activity: Median Filtering in ImageJ

Start ImageJ and load the image dusty.bmp from the CD. You will see that the image has a lot of dust on it (it was digitized from a dirty 35-mm slide). The appearance of the image can be improved using the median filter.

Select Process | Filters | Median. A dialog window (Figure 3.7) will appear. Choose a radius of 1 or 2 pixels, then click on the OK button. The image should now look a lot less dusty.

Load any new image of your choice.

Select Process | Noise | Salt and Pepper to add noise from the extremes of the gray range to this image.

Repeat if desired to get more dots.

Select Process | Filters | Median. Choose a radius of 1 pixel, then click on the OK button. Providing you did not add too many dots, this operation should have removed them. The median filter works well on impulse noise because the noise pixels are always at the extremes of the ranked list and will be replaced by the median value, which can be assumed to be representative of the underlying image and not noise.

Load the image PET_Scan_256.tif from the CD. This is also a noisy image, but in this case the noise pattern does not consist of single pixels of impulse noise. It is not expected that the effect of median filtering will be so impressive here, because, in contrast to the situation with impulse noise, the median pixel in the ranked list may itself contribute to the noise. Use ImageJ to apply median filtering to this image. The image will be blurred by the median filter, and may indeed look a little smoother, but the effect is much less striking than the effect on impulse noise. Convolution or hybrid filtering strategies are commonly employed to reduce non-impulse image noise, and these are covered later in this chapter.

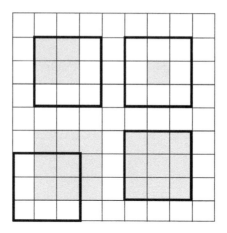

**FIGURE 3.8**

Median filtering. In each of the four examples, the 3×3 neighborhood is indicated by the black square overlying the image. The pixel value centered under the 3×3 neighborhood will be changed from gray to a white background pixel in all but the bottom right example, because fewer than half of the pixels in the neighborhood share a similar gray value to the pixel of interest.

### 3.2.1.1   Effect of Neighborhood Size and Shape

Larger filters change the image more, because pixels from further away from the pixel of interest are included in the calculations. Additionally, the nature of the median filter means that if fewer than half of the pixels in the neighborhood share a similar gray value to the pixel of interest, then the feature they represent will be removed from the image (Figure 3.8). The effect is also mediated by the shape chosen for the neighborhood, as this can determine whether or not half of the pixels in the neighborhood share a similar gray value to the pixel of interest.

The first six shapes and sizes of rank filtering neighborhoods available in the program ImageJ are shown in Figure 3.9.

*Self-assessment question 3.03*

Consider the six neighborhoods shown in Figure 3.10a–f. Each neighborhood is shown outlined in black and overlying a binary image, which is a triangular object on a background. The pixel of interest is indicated by a black dot.

A. Neighborhood (a) has 9 pixels, which other neighborhood also has 9 pixels?

B. Which neighborhood is the same size as neighborhood (c), but has a different shape?

C. In (a) the pixel of interest will be replaced, for median filtering, with a pixel with the background value, because the neighborhood contains

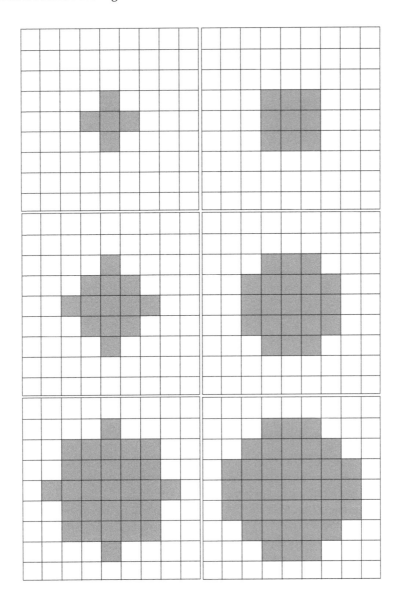

**FIGURE 3.9**
The first six shapes and sizes for rank filtering neighborhoods in the program ImageJ, indicated by the shaded shapes. Selection of a different shape can change the filtering result, as the shape can affect whether or not half of the pixels in the neighborhood share a similar gray value to the pixel of interest.

five background pixels and only four object pixels. For (b)–(f), will the pixel of interest be an object or a background pixel following median filtering?

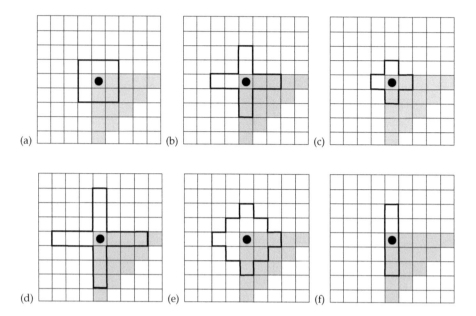

**FIGURE 3.10**
(a)–(f) Images for self-assessment question 3.03. Six neighborhoods shown outlined in black and overlying a binary image, which is a triangular object on a background. The pixel of interest is indicated by a black dot.

### 3.2.2 Convolution Filtering

Like rank filtering, convolution filtering is concerned with replacing a pixel value with a new value found from the values of its neighbors, but there are a few more calculations involved. A square kernel is defined to represent the neighborhood around the pixel. Each location in the kernel is assigned a numerical value, known as a weight. In Figure 3.11a, these weights are given by the letters a to i. The kernel is moved across the image, and the pixel value under the center of the kernel is replaced by the weighted sum of the surrounding pixels. In the example of a section of image in Figure 3.11b, where the pixel values are represented by upper case letters, the pixel $M$ will be replaced by $M'$, where

$$M' = aG + bH + cI + dL + eM + fN + gQ + hR + iS \qquad (3.1)$$

This summation of the results of a series of multiplications is closely related to the mathematical operation called convolution, which is indicated using the symbol "*" or a star in a circle, and this is why these filters are called convolution filters. Convolution filtering may also be called linear filtering.

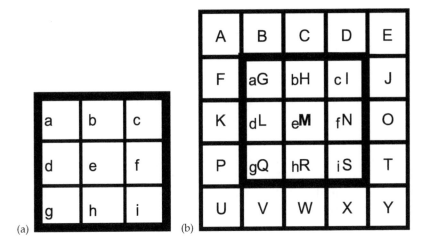

(a)                    (b)

**FIGURE 3.11**
Convolution filtering. (a) The weights in a 3×3 convolution kernel are indicated by the letters a to i, (b) the 3×3 convolution kernel overlying a section of an image, with pixel values indicated by upper-case letters. The pixel value M will be replaced by M′ as indicated in Equation (3.1).

### Convolution and Correlation

The operation illustrated in Figure 3.11 is strictly the mathematical correlation operation. The convolution operation is the same but with the kernel rotated through 180°. However, as many of the kernels used in image processing are rotationally symmetric, the results from correlation and convolution come out the same. Indeed, in some image processing programs it is assumed that all kernels are rotationally symmetrical, and convolution filtering is actually implemented using the correlation operation. It is worth checking to be sure what a program is doing.

*Self-assessment question 3.04*

What would be the effect of using the 3×3 convolution kernel:

$$h = \begin{bmatrix} 0 & 0 & 0 \\ 0 & 1 & 0 \\ 0 & 0 & 0 \end{bmatrix} \tag{3.2}$$

Answer this without using a calculator.

Commonly available convolution filters are the average or mean filters, and Gaussian filters used to smooth and blur the image, the sharpening filter and edge enhancement filters.

The averaging filter is used to reduce noise in an image, and it will also cause blurring. This is an appropriate filter to use when the noise is not limited to extreme values, unlike the impulse noise that was successfully removed using the median filter. A typical averaging kernel is given in Equation (3.3).

$$h = \begin{bmatrix} 1/9 & 1/9 & 1/9 \\ 1/9 & 1/9 & 1/9 \\ 1/9 & 1/9 & 1/9 \end{bmatrix} \quad \text{or} \quad h = 1/9 \begin{bmatrix} 1 & 1 & 1 \\ 1 & 1 & 1 \\ 1 & 1 & 1 \end{bmatrix} \tag{3.3}$$

The kernel weights are normalized so that their sum is 1 and the overall brightness of the image is left unchanged in the result. The result of applying this averaging filter is shown in Figure 3.12b, and the result for a similar but larger 7x7 kernel in Figure 3.12c. The Gaussian filter also has a blurring effect, but the weights at the edges of the kernel are smaller than those in the center to reduce the effect of pixels further away from the pixel of interest. The result of applying the 3×3 Gaussian filter of Equation (3.4) is shown in Figure 3.12d.

$$h = \begin{bmatrix} 0.2098 & 0.458 & 0.2098 \\ 0.458 & 1 & 0.458 \\ 0.2098 & 0.458 & 0.2098 \end{bmatrix} \tag{3.4}$$

The sharpening filter increases gray level contrast and accentuates detail in an image. It will also accentuate noise. The sharpening filter in ImageJ uses the kernel given in Equation (3.5). The result of applying this 3×3 sharpening filter is shown in Figure 3.12e.

$$h = \begin{bmatrix} -1 & -1 & -1 \\ -1 & 12 & -1 \\ -1 & -1 & -1 \end{bmatrix} \tag{3.5}$$

For performing edge enhancement, the kernel weights are chosen to give a different result depending on the direction of the underlying edge in the image. For example, the kernels in Equations (3.6) and (3.7) emphasize edges that run vertically and horizontally, respectively. Results on a test image are shown in Figure 3.13.

$$h_1 = \begin{bmatrix} -1 & 0 & 1 \\ -1 & 0 & 1 \\ -1 & 0 & 1 \end{bmatrix} \tag{3.6}$$

$$h_2 = \begin{bmatrix} -1 & -1 & -1 \\ 0 & 0 & 0 \\ 1 & 1 & 1 \end{bmatrix} \tag{3.7}$$

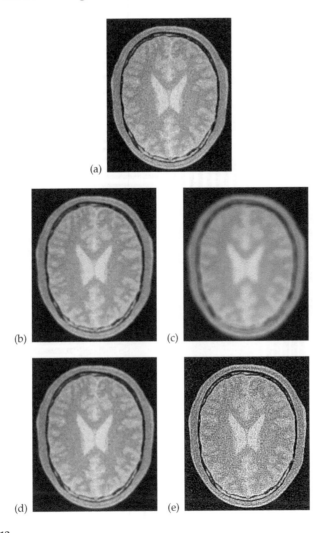

**FIGURE 3.12**
Results of applying various convolution filters. (a) Original noisy image, (b) result of applying a 3×3 averaging filter, (c) result of applying a 7×7 averaging filter, (d) result of applying a 3×3 Gaussian filter, (e) result of applying a 3×3 sharpening filter. (Image data from the BrainWeb Simulated Brain Database [1,2]. With permission.)

*Self-assessment question 3.05*

Figure 3.14 shows the pixel values for a section of an image with a 3x3 convolution kernel overlaid at the top left. Find the new value for the central pixel under the kernel, currently with value 89, for each of the convolution kernels $h_a$ and $h_b$ in Equation (3.8).

$$h_a = \begin{bmatrix} 1/9 & 1/9 & 1/9 \\ 1/9 & 1/9 & 1/9 \\ 1/9 & 1/9 & 1/9 \end{bmatrix} \qquad h_b = \begin{bmatrix} -1 & -1 & -1 \\ 0 & 0 & 0 \\ 1 & 1 & 1 \end{bmatrix} \qquad (3.8)$$

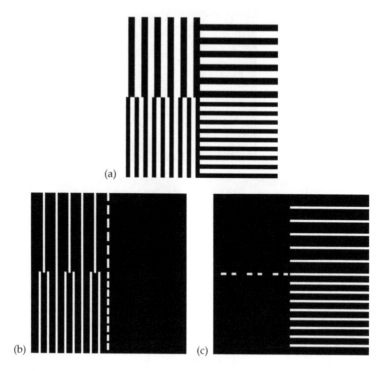

(a)

(b)          (c)

**FIGURE 3.13**

Edge enhancement by convolution filtering. (a) Original image, (b) result of applying kernel in Equation (3.6), (c) result of applying kernel in Equation (3.7).

| 98 | 112 | 134 | 154 | 169 |
|----|-----|-----|-----|-----|
| 89 | 89 | 169 | 169 | 169 |
| 89 | 89 | 89 | 169 | 169 |
| 89 | 89 | 89 | 89 | 89 |
| 89 | 89 | 89 | 89 | 89 |

**FIGURE 3.14**

Image for self-assessment question 3.05, showing the pixel values for a 5×5 section of an image, with a 3×3 convolution kernel outlined in black at the top left of the section.

**Activity: Convolution Filtering in ImageJ**

Start ImageJ and load the image tip.bmp from the CD. Repeat, so that you have two windows labeled tip.bmp.

Click on one of the two images to select it, then select Process | Smooth.

Visually compare the smoothed image with the unprocessed version in the other window.

Use the Edit | Undo function to undo the smoothing operation, watching closely to see how the image changes. You may not see very much as the smoothing filter is not a very strong filter. To get a stronger effect, perform the operation repeatedly on the same image, and you should eventually be able to see the difference.

Close the open images.

Load the image tip.bmp from the CD. Use Image | Duplicate to create another copy of it. Click on one of the two images to select it, then select Process | Find Edges. This convolution filtering operation highlights sharp changes in gray level in the image.

### More Convolution Filtering in ImageJ

The kernels used for the Smoothing and Find Edges operations are described in the Process Menu section of the ImageJ Documentation. It is also possible to filter using a custom convolution kernel. This option is also described in the Process Menu section of the ImageJ Documentation and is included as an activity in Chapter 4.

### 3.2.2.1   *Effect of Kernel Size*

Larger filters change the image more, because pixels from further away from the pixel of interest are included in the calculations. Images produced using large convolution kernels are generally not helpful for visual assessment, but large kernels may be useful as part of a series of operations.

**Activity: Processing Time for Different Convolution Kernel Sizes in ImageJ**

Start ImageJ and select File | New | Image ... .

Set values for a single 512×512 8-bit image filled with a ramp.

Select Process | Filters | Gaussian Blur ... and set the radius to 1 pixel and click on OK.

Make a note of the time taken for the operation, which is shown at the bottom left of the ImageJ menu (e.g., 0.157 seconds).

Select Edit | Undo.

Select Process | Filters | Gaussian Blur ... and set the radius to 128 pixels and click on OK.

Make a note of the time taken for the operation, which is shown at the bottom left of the ImageJ menu (e.g., 2.375 seconds).

Close open images and exit from ImageJ.

### 3.2.3   Hybrid Filtering

Hybrid filters may include both rank and convolution steps, or may involve an additional image processing operation. A hybrid filter that performs a rank filtering operation and a convolution smoothing operation is a good choice where both impulse and Gaussian noise are present. A widely used hybrid filter is the *unsharp mask*. This is a filter that sharpens and enhances edges, and it involves the subtraction of a weighted copy of a smoothed version of the image from the original image. Both the degree of smoothing and the value of the weighting affect the result (Figure 3.15). The unsharp mask filter can be found in the Process | Filters menu in ImageJ.

## 3.3   Adaptive Filtering

Adaptive filters perform a different operation depending on the image content in the region in which they are being applied. The filters discussed in this chapter were assumed to perform in exactly the same way across the whole image, but an adaptive filter can be designed, for example, to blur only those parts of an image that do not contain edges.

## 3.4   Good Practice Considerations

In common with other image processing operations, when used as part of a quantitative protocol, spatial domain filters should not be applied in an ad hoc manner:

- The justification for the choice of kernel size, neighborhood size and shape, and values of other parameters should be recorded and reported.
- The filtering should be applied consistently to images that are to be compared.

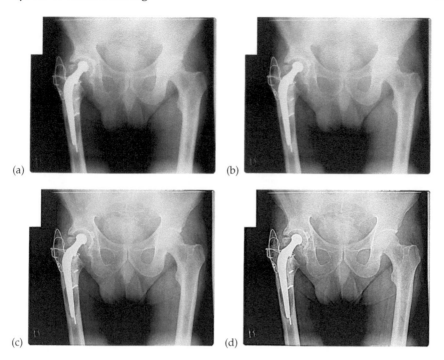

(a)  (b)  (c)  (d)

**FIGURE 3.15**

Effect of the choice of parameters on unsharp mask filtering. (a) Original image, (b) result of unsharp mask filtering with smoothing radius 2 and mask weight 0.6, (c) result of unsharp mask filtering with smoothing radius 10 and mask weight 0.6, (d) result of unsharp mask filtering with smoothing radius 10 and mask weight 0.8.

Observers should always be told what filtering has been applied and have access to the original image data, even for qualitative assessment. This is because there are pitfalls when the results of image processing mimic clinical conditions.

## 3.5 Chapter Summary

Spatial domain filtering was covered in this chapter. The emphasis was on rank and convolution filtering, and examples were given of the key convolution kernels. Hybrid methods and adaptive filtering were also described briefly. ImageJ activities were included to allow the reader to perform filtering using some of the spatial domain filters built into the program.

## 3.6 Feedback on the Self-Assessment Questions

### Self-assessment question 3.01

A transverse T2-weighted magnetic resonance (MR) image of the same slice would not have a strong signal from the bone of the skull, whereas the skull is clearly shown on the computed tomography (CT) image in Figure 3.1. The MR image would show more detail within the soft tissue of the brain, with gray matter and white matter differentiated. Any fluid-filled areas would have a high signal in the MR image.

**Self-assessment question 3.02**

The results of the maximum filter for a 3×3 square neighborhood:
a) 169, b) 169, c) 169, d) 89.

The results of the minimum filter for a 3×3 square neighborhood:
a) 89, b) 89, c) 89, d) 89.

**Self-assessment question 3.03**

- A. Neighborhood (b) also has 9 pixels.
- B. Neighborhood (c) has 5 pixels; the other neighborhood of the same size is (f).
- C. After median filtering, the pixel of interest will be an object pixel for (b), (c), (d), and (f) and a background pixel for (e). From these examples, it can be seen that the shape of the neighborhood is important as well as its size.

**Self-assessment question 3.04**

This kernel would not change the image at all, as the only pixel value not multiplied by zero is the pixel of interest.

**Self-assessment question 3.05**

Convolution kernel $h_a$ is an averaging kernel and the value 89 would be replaced by 106.

Convolution kernel $h_b$ is an edge detecting kernel and the value 89 would be replaced by –77.

---

## References

1. The BrainWeb Simulated Brain Database. www.bic.mni.mcgill.ca/brainweb/
2. Collins, D.L. et. al., Design and construction of a realistic digital brain phantom, *IEEE Trans. Med. Imag.*, 17, 463, 1998.

# 4

## Frequency Domain Filtering

## 4.1 Introduction

Frequency domain filtering has an effect equivalent to convolution filtering, which was covered in Chapter 3. The way in which the result is achieved is, however, different. Like convolution filtering, the operations fall into the image enhancement category, and the results may also be used for preprocessing in image analysis protocols. With frequency domain processing, it is easier to understand the relation between the filter parameters and their effect on the image than when using spatial domain methods, and this means that it is a good choice for quantitative image analysis. Furthermore, the time taken for computation is less for all but the simplest of operations.

### 4.1.1 Learning Objectives

When you have completed this chapter, you should be able to

- Recognize when images have undergone low-pass or high-pass filtering
- Understand the relationship between filtering in the spatial and frequency domains
- List the advantages of filtering in the frequency domain
- Recognize different applications of the word "convolution"
- Use ImageJ to perform frequency domain filtering

### 4.1.2 Example of an Image Used in this Chapter

*Self-assessment question 4.01*

Which part of the body appears in the image in Figure 4.1? Which imaging modality has been used?

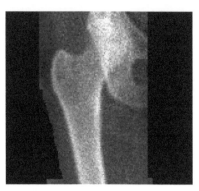

**FIGURE 4.1**
Image for self-assessment question 4.01.

### 4.1.3   Convolution Filtering: A Reminder

In Chapter 3, several convolution filters were introduced. The important points are summarized in the following list:

- Convolution is a mathematical operation, indicated using the symbol "*" or a star in a circle.

- Convolution involves the summation of the results of a series of multiplications

- In convolution filtering, the convolution kernel defines the factors by which pixel values are multiplied

- Examples of convolution filters include the sharpening kernel and the averaging kernel

In this chapter, an alternative way of performing convolution filtering is introduced. To understand this, it is necessary first to spend a little time getting reacquainted with the idea of spatial frequency, which was covered in Chapter 1.

## 4.2   The Spatial Domain and the Frequency Domain

Shapes in images are made up of changes in gray level across the image, from dark to light to dark. The rate of this change is called the spatial frequency. In each of the images in Figure 4.2, there is just one spatial frequency. That spatial frequency is different in each image.

*Self-assessment question 4.02*

Put the images (a), (b) and (c) of Figure 4.2 in order of increasing spatial frequency, starting with the image with the lowest spatial frequency

**FIGURE 4.2**

(a)–(c) Images for self-assessment question 4.02, illustrating the concept of spatial frequency in images. Each image contains a different, single spatial frequency.

A real image will be much more complex and will contain many different spatial frequencies. However, we can say that:

- High spatial frequencies correspond with fine detail, such as noise and edge features.
- Low spatial frequencies correspond with larger objects that have a fairly uniform gray level within the object.

These two properties mean that it is possible to describe the spatial frequency content of an image in very broad terms, from observation of the image. This is illustrated in self-assessment question 4.03.

*Self-assessment question 4.03*

Figure 4.3a has a slightly noisy appearance, with square pixels visible in some areas. The appearance of the noise has been reduced in Figure 4.3b by applying a 7×7 Gaussian convolution filter. Using knowledge about which features are represented by low and high spatial frequencies, would you expect the filtered image (b) to have:

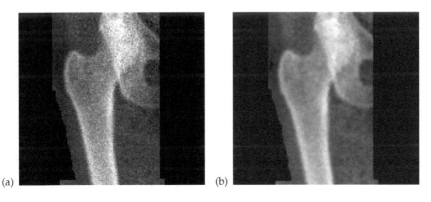

**FIGURE 4.3**

Images for self-assessment question 4.03. (a) Original image. (b) Image after application of a 7×7 Gaussian convolution filter.

- more high spatial frequencies than the unprocessed image?
- fewer high spatial frequencies than the unprocessed image?
- the same quantity of high spatial frequencies as the unprocessed image?

---

## 4.3   Frequency Domain Filtering

This spatial frequency viewpoint leads to an alternative way of considering the filtering process, one where filtering is considered in terms of the spatial frequencies. The filter that was used in self-assessment question 4.03 was one that preserves only the low spatial frequencies in an image. This type of filter is often described as a *low-pass* filter. Naturally there is also such a thing as a *high-pass* filter: one that preserves high spatial frequencies but removes the low. A *band-pass* or notch filter preserves a particular band, or range, of frequencies. One can imagine the high-pass filter to act like a garden riddle or a kitchen sieve. All the fine details (i.e., high spatial frequencies) get through the mesh, but the larger features cannot pass.

*Self-assessment question 4.04*

Which of the two filtered images in Figure 4.4, (b) or (c), represents the result of applying a high-pass filter to the original image in Figure 4.4a?

Clearly some of the spatial domain filters introduced in Chapter 3 can be considered in terms of their effect on the spatial frequencies in the image. It is also possible to generate the same effects that can be achieved by spatial domain filtering, but by working directly with the spatial frequencies. In general terms, the procedure is as follows:

- Generate a representation to show which spatial frequencies are present in the image
- Remove selected spatial frequencies from the representation

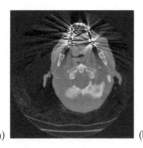 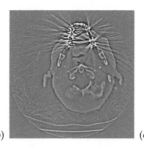 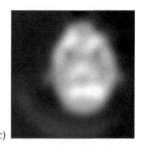

(a)    (b)    (c)

**FIGURE 4.4**

Images for self-assessment question 4.04. (a) Original image. (b)–(c) Images after application of different frequency domain filters. (Image data from the Chapel Hill Volume Rendering Test Data Set, Volume I. [1] With permission.)

- Reverse the process to get back to an image. The new image is filtered because some spatial frequencies necessary to reproduce the original image are no longer present.

A standard mathematical technique can be used to implement the procedure. The representation showing which spatial frequencies are present in the image is called a *spectrum,* and it is obtained using Fourier transformation. Specific spatial frequencies are removed from the spectrum by filtering it; the mechanism for filtering will be described later. To distinguish this method from filtering in the spatial domain, the process is called filtering in the frequency domain, or in frequency space. Finally, the Inverse Fourier transform is applied to the filtered spectrum to give the filtered image.

### Fourier Analysis

Those with a mathematical background will know that Fourier theory states that any waveform can be expressed as a number of sine and cosine terms, each with a different frequency, amplitude and phase. Fourier analysis breaks down a signal into these component functions. If the waveform is rebuilt, but some of the frequency components are omitted, then the result differs from the original waveform. For example, if the high-frequency components are omitted, then edges and corners are not well represented in the result.

## 4.3.1 Advantages of Filtering in the Frequency Domain

Why filter in the frequency domain, when spatial domain convolution filtering works so well? There are three situations in which it is preferred over spatial domain convolution filtering:

- It is possible to perform operations that are difficult in the spatial domain, such as removing, or enhancing, only specific frequencies in the image.
- Periodic patterns can be selectively removed or enhanced (e.g., scanning lines).
- For operations that require large kernel sizes in the spatial domain, frequency domain filtering is computationally faster.

These points are considered in more detail in the following sections.

## 4.3.2 The FFT and the Spectrum

The Fourier transform, and its inverse, can be applied using many image processing packages, including ImageJ. In most cases, a particular method of calculating the Fourier transform is used. This is the Fast Fourier Transform,

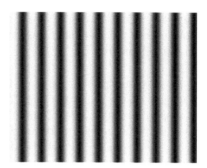

**FIGURE 4.5**
An image containing a single dominant spatial frequency.

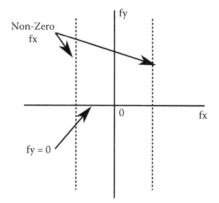

**FIGURE 4.6**
Sketch of the spectrum expected from Fourier transformation of the image in Figure 4.5.

which is abbreviated to FFT. Some examples follow of spectra produced by applying the FFT.

One of the simplest images to consider is one with a single dominant spatial frequency, such as the striped image of Figure 4.5. In this image, the following points may be noted:

- The stripes are vertical stripes. This means that the changes in intensity run in the horizontal (x) direction.
- There is one dominant spatial frequency present in the x direction, with a non-zero value.
- In the y direction, the intensity is uniform, so the frequency of intensity changes is zero. That is, the dominant frequency in the y direction has a frequency of zero.

From these observations, it is possible roughly to sketch the form of the spectrum (Figure 4.6). Note that

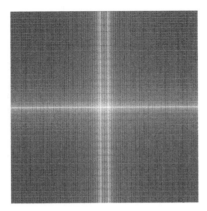

**FIGURE 4.7**
Spectrum obtained from Fourier transformation of the image in Figure 4.5.

- The spectrum includes both positive and negative frequencies for both x and y. This means that although there is only one dominant frequency in the x direction, the graph contains two lines, one each for both positive and negative values.
- The origin of the graph (0,0) is at the center.

When the FFT is applied to the striped image of Figure 4.5 in ImageJ, the result is as shown in Figure 4.7.

*Self-assessment question 4.05*
Sketch your own copy of the result of the FFT (Figure 4.7), and identify on it:

- The feature that represents the zero frequency y component
- The feature that represents the dominant x component

The next stage in frequency domain filtering is to ensure that areas of the spectrum that are not required are set to zero (black in the examples here) before an inverse FFT is performed (Figure 4.8).

*Self-assessment question 4.06*
What do you expect to see in the final image if the spectrum in Figure 4.7 is altered to appear as in Figure 4.8? Will the image still contain stripes? Will the noise in the image be increased or decreased?

In the example for Figure 4.8, the pattern used to set parts of the spectrum to zero was not symmetrical in the x and y directions. Higher x frequencies were blocked, but some high y frequencies were not. More typically, the distribution of areas set to zero (called the *frequency domain filter*) is often symmetrical in the x and y directions.

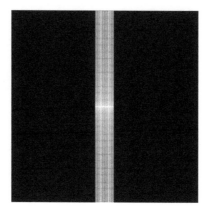

**FIGURE 4.8**
A filtered version of the spectrum in Figure 4.7. Areas representing frequencies that are not desired in the filtered image have been set to zero (black).

Two examples of frequency domain filters are shown in Figure 4.9b–c. In both cases the filter is symmetrical about the origin of the spectrum, meaning that the x and y frequencies are affected in the same way. As before, black areas set the spectrum to zero while the spectrum is unchanged by white.

*Self-assessment question 4.07*

The two example filters in Figure 4.9b and Figure 4.9c were applied to the FFT of the head image shown in Figure 4.9a. After inverse transformation, the results in Figure 4.9d and e were obtained. Which of the two filters (b) and (c) goes with which output image (d) and (e)?

### 4.3.3 Application of the Filter to the Spectrum

Filtering is achieved by setting unwanted parts of the spectrum to zero, while leaving other areas unchanged. In the filters shown in Figure 4.9, black has a value of 0 and white has value 1. If a spectrum were multiplied by such a filter, then the desired effect will be achieved.

In summary, the steps involved to filter in the frequency domain are:

- Calculate the FFT of the image.
- Design a frequency domain filter to be applied to that FFT. This has value 0 for the spatial frequencies to be removed, 1 for those to be retained.
- Multiply the FFT of the image by the frequency domain filter.
- Calculate the inverse FFT of the filtered spectrum.

The discussion so far has concentrated on the removal of specific features in the spectrum, but it is also possible selectively to enhance features, or to reduce their effect without removing them completely.

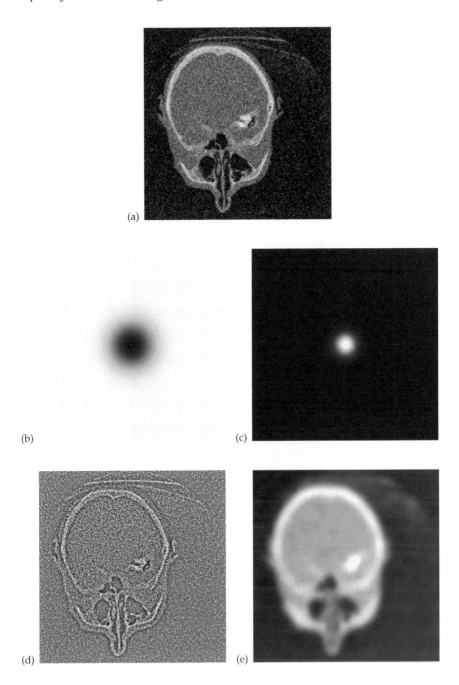

**FIGURE 4.9**
Frequency domain filtering. (a) Original image, (b)–(c) examples of frequency domain filters, (d)–(e) images for self-assessment question 4.07, obtained by filtering the spectrum of the original image with the frequency domain filters. (Image data from the Chapel Hill Volume Rendering Test Data Set, Volume I. [1] With permission.)

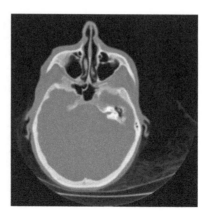

**FIGURE 4.10**
The image CThead.tif, which may be found on the CD. (Image data from the Chapel Hill Volume Rendering Test Data Set, Volume I. [1] With permission.)

**FIGURE 4.11**
The result of applying the FFT operation to the image CThead.tif.

### Activity: Low-pass Frequency Domain Filtering Using ImageJ

Select File | Open and load the image CThead.tif* from the CD (Figure 4.10).

Select Process | FFT | FFT. The result should appear as Figure 4.11.

Select the Rectangular selections option from the ImageJ tools (Figure 4.12) and draw a square on the FFT, centered on the zero frequency point at the center of the image (Figure 4.13).

Low-pass filtering is required. This means that the high spatial frequency areas of the spectrum should be set to zero. Select Edit | Clear Outside (Figure 4.14)

---

* Image data from the Chapel Hill Volume Rendering Test Data Set, Volume I. [1] With permission.

**FIGURE 4.12**
The ImageJ rectangular selections icon in the ImageJ Tools menu.

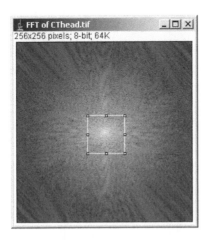

**FIGURE 4.13**
A rectangular selection drawn on the spectrum of the original image.

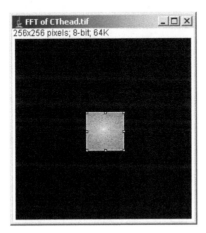

**FIGURE 4.14**
The spectrum edited so that high spatial frequency areas (outside the selected square region of interest) are set to zero.

Select Edit | Selection | Select None (this simply removes the square that was drawn).

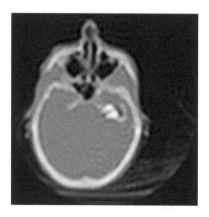

**FIGURE 4.15**
The result of applying the inverse FFT operation to the filtered spectrum shown in Figure 4.14. The original image has undergone low-pass filtering. (Image data from the Chapel Hill Volume Rendering Test Data Set, Volume I. [1] With permission.)

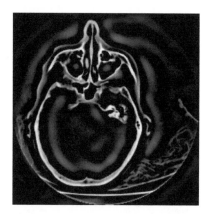

**FIGURE 4.16**
An example of the result expected for the activity on High-Pass Frequency Domain Filtering Using ImageJ; this is a high-pass filtered image. The image is shown here with additional contrast enhancement. (Image data from the Chapel Hill Volume Rendering Test Data Set, Volume I. [1] With permission.)

Select Process | FFT | Inverse FFT (Figure 4.15). This displays the low-pass filtered result. When compared with the original image, the image has less detail and looks blurred. The wavy patterns arise because of the sharp edges of the filter that was used.

### Activity: High-Pass Frequency Domain Filtering Using ImageJ

Repeat the last Activity, but use Edit | Clear instead of Edit | Clear Outside, so that high-pass filtering is performed. A typical result, using a much smaller square than for the low-pass filter, is shown in Figure 4.16.

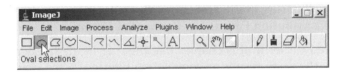

**FIGURE 4.17**
The ImageJ elliptical selections icon in the ImageJ Tools menu.

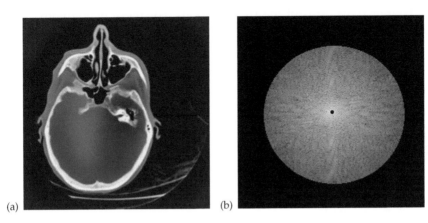

(a)          (b)

**FIGURE 4.18**
(a) The image CThead.tif after band-pass filtering, (b) the edited spectrum associated with the band-pass filtered image. (Image data from the Chapel Hill Volume Rendering Test Data Set, Volume I. [1] With permission.)

## Activity: Band-Pass Frequency Domain Filtering Using ImageJ

Repeat the activity again, this time performing band-pass filtering by centering a small circle on the spectrum and using Clear to set the inside to zero. Then center a larger circle and use Clear Outside to set the outermost frequencies to zero. Remember that you will need to change from Rectangular Selections to Elliptical Selections in order to draw a circle (Figure 4.17).

The result in Figure 4.18a was obtained by editing the spectrum as shown in Figure 4.18b. Image | Adjust | Brightness/Contrast was used to make the result display more clearly.

*Self-assessment question 4.08*

A real example of the usefulness of frequency domain filtering is described at http://rsb.info.nih.gov/ij/docs/examples/FFT/. Frequency domain filtering was used to remove horizontal lines on a scanning laser confocal microscope image. Adapt the technique used in the earlier activities to remove the horizontal lines from the image confocalimage.gif,* which is on the CD.

---

* From http://rsb.info.nih.gov/ij/docs/examples/FFT/. With permission.

### 4.3.4 Why Filtering in the Frequency Domain Is the Same as Convolution Filtering in the Spatial Domain

It was hinted earlier in this chapter that filtering in the frequency domain would give the same results as convolution filtering in the spatial domain. The results of the two methods are indeed the same, and this can be explained using the *convolution theorem*. The convolution theorem states that the result of the convolution operation between two functions in the spatial domain is equal to the inverse Fourier transform of the result of multiplying together the Fourier transforms of the two functions. For those happy with mathematics, this is expressed using the mathematical notation below.

#### The Convolution Theorem

The convolution theorem can be expressed as

$$f * h \Leftrightarrow FH$$

The left-hand side of the expression represents the convolution of an image, $f$, with a kernel $h$. $F$ is the Fourier transform of the image and $H$ is the Fourier transform of the kernel, so the right-hand side of the expression is the product of these two. The symbol $\Leftrightarrow$ indicates Fourier transformation from left to right, and inverse Fourier transformation from right to left.

At first glance, because of the need to perform the Fourier transforms, it looks as if performing frequency domain filtering is a lot more complicated than the equivalent operation in the spatial domain. However, convolution involves many multiplication operations, and the number of multiplications gets bigger as the size of the kernel is increased. In contrast, although two forward transforms and one inverse transform are required for the frequency domain operation, these take the same time to compute whatever filter is in use. Furthermore, there is only the one multiplication to be performed. The reduction in the number of multiplications means that filtering in the frequency domain is computationally faster than filtering in the spatial domain, in spite of the need to perform the Fourier transforms.

The images in Figure 4.19 demonstrate how the results of filtering are the same via the two alternative routes. Figure 4.19a shows the original image, and the two filtered images are in Figure 4.19b and Figure 4.19c.

#### Activity: Comparison of Frequency Domain and Convolution Filtering Using ImageJ

This activity also introduces two useful features in ImageJ: the ability to perform convolution filtering with a kernel of one's own design, and the capacity for frequency domain calculations.

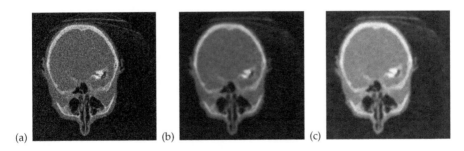

(a) (b) (c)

**FIGURE 4.19**
A comparison of filtering in the spatial and frequency domains. (a) Original image. (b) Result of spatial domain filtering. (c) Result of frequency domain filtering. (Image data from the Chapel Hill Volume Rendering Test Data Set, Volume I. [1] With permission.)

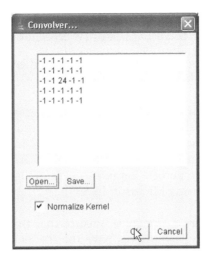

**FIGURE 4.20**
The ImageJ Convolver dialog for convolution filtering.

Select File | Open and load the image CThead.tif* from the CD.

Select Image | Type | 32-bit. This converts the image to the 32-bit data type, which is necessary correctly to cope with negative pixel values.

Select Process | Filters | Convolve. The ImageJ dialog window entitled "Convolver" will open. The default convolution kernel will probably be displayed (Figure 4.20). If it is not, type the kernel in (it consists of five rows and five columns. All the values are –1 except the central value, which is 24).

---

* Image data from the Chapel Hill Volume Rendering Test Data Set, Volume I. [1] With permission.

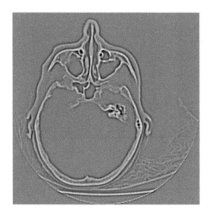

**FIGURE 4.21**
The result of convolution with a 5×5 edge-enhancement kernel. (Image data from the Chapel Hill Volume Rendering Test Data Set, Volume I. [1] With permission.)

Ensure there is a tick beside "Normalize Kernel," then click OK. The result of filtering is an edge-enhanced image (Figure 4.21).

It was emphasized earlier that convolution filtering is computationally slow, but the preceding operation was probably achieved almost instantly. This is because a 5×5 kernel is a relatively small one. In the next few steps of the activity, the 5×5 kernel will be made larger by using ImageJ to surround the central values with zeroes. Large kernels can also be generated outside ImageJ, for example, using Microsoft Excel®*.

Select Process | Filters | Convolve. Ensure that the correct kernel is displayed, then click the Save button. Save the file as defaultkernel.txt. Click Cancel to exit the Convolver window.

Select File | Import | Text Image and import defaultkernel.txt.

Select Image | Adjust | Canvas Size. Set the Width and Height to 63, Position to Center and place a tick next to Zero Fill. Click OK.

Select File | Save As | Text Image and save as defaultkernel63x63.txt. Close the kernel image.

Load a new copy of CThead.tif from the CD and convert it to a 32-bit image.

Select Process | Filters | Convolve to apply defaultkernel63x63.txt – use the Open button to open the file and import the numbers into the box, rather than accepting the default kernel or typing in values.

---

* Registered trademark of Microsoft Corporation, Redmond, Washington, USA.

This time, the convolution processing takes much longer. The word "Convolve" will be shown at the bottom left of the ImageJ window, and a thermometer display will appear at the right. When the processing is complete, the number of seconds taken will be displayed. The resulting image should be the same as before, because the same non-zero values are at the center of the kernel. The point of the exercise was to show how the size of the kernel affected the processing time. If the resulting image is black and white, with no grays, it is likely that the conversion of the original image to 32 bits was omitted.

In the next steps the same filtering is done in the frequency domain.

Close any open images.

Load CThead.tif from the CD.

Import the file defaultkernel.txt as a Text Image.

Set the Image Type for CThead.tif to 32-bit.

For the FD Math operation that follows, the image sizes must match, so change the canvas size for defaultkernel.txt to 256×256, with the Center and Zero Fill options selected.

Select Process | FFT | FD Math. The ImageJ dialog window entitled "FFT Math" will appear (Figure 4.22). Ensure that the two different image names appear for Image1 and Image2, set the operation to "Convolve," leave the name "Result" in the result box and ensure that there is a tick beside "Do Inverse Transform." Click on OK.

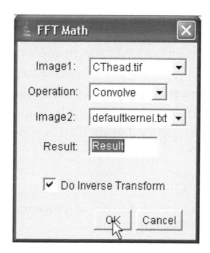

**FIGURE 4.22**
The ImageJ frequency domain mathematics dialog.

ImageJ now performs an FFT on both the images, multiplies these transformed images together and performs an inverse transform. In other words, it performs frequency domain filtering using the same kernel as was used for convolution filtering. The result should be visually identical to those generated in previous steps. If not, check that the change from 8- to 32-bit data type was performed.

Note that the operation was performed quickly. Frequency domain filtering is much faster than spatial convolution when the convolution kernel is larger than a few values.

### 4.3.5   Use of the Word Convolution

The word convolution refers to a specific mathematical operation (Chapter 3). It is true that one method of image enhancement is convolution filtering, but remember that not all spatial domain image enhancement filters are convolution filters. For example, the median filter is a rank filter, which does not involve the convolution operation. Sometimes, rather loosely, people will use the term image convolution to refer to all image enhancement operations. This is incorrect, because convolution is not a synonym for image enhancement.

The other reason for taking care with the word is that confusion can arise when discussing *image deconvolution*. Deconvolution is, as you might expect, a reverse convolution operation. However, in image processing it has a particular meaning: deconvolution is a process that is designed to remove the specific effects of the imaging process from an acquired image or signal. Deconvolution falls into the image restoration category of image processing, and it is considered alongside other methods of image restoration in Chapter 7.

*Self-assessment question 4.09*

Which of the following five options correctly describes a step in frequency domain filtering?

    a) The FFT of the image is added to the frequency domain filter

    b) The FFT of the image is subtracted from the frequency domain filter

    c) The FFT of the image is multiplied by the frequency domain filter

    d) The FFT of the frequency domain filter is divided by the image

    e) The FFT of the frequency domain filter is added to the FFT of the image

*Self-assessment question 4.10*

Which of the following are advantages of frequency domain filtering over spatial domain convolution filtering?

    a) Ability selectively to remove periodic patterns in an image

    b) Ability selectively to enhance periodic patterns in an image

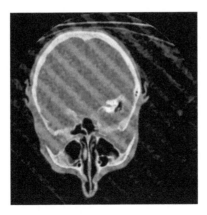

**FIGURE 4.23**
Image with stripes for self-assessment question 4.11. (Image data from the Chapel Hill Volume Rendering Test Data Set, Volume I. [1] With permission.)

c) Faster to compute when large kernels would be involved

d) Ability selectively to remove details of a particular size from the image

e) Ability selectively to enhance details of a particular size from the image

*Self-assessment question 4.11*

A copy of the image in Figure 4.23 is on the CD, named ctandstripes05.tif. Use ImageJ to produce a new image which has the stripes removed, and note the key steps in the processing that was undertaken to produce the final image.

## 4.4   Good Practice Considerations

Exactly the same considerations apply to frequency domain filtering as to spatial domain operations (Chapter 3). In particular, there is a need to justify both the use of a particular type of filter and the choice of parameters. Examples of the kind of question to address are given here:

- Why was this particular range of frequencies chosen for filtering? This choice will probably be related to the scale of an image feature to be removed or enhanced, and knowledge of the spatial resolution of the images must be included.

- Have steps been taken to ensure that the choice of parameters is equally valid for all images processed? It may be necessary to describe the scale of a feature as a proportion of the size of a large feature, and not in absolute terms.

Always be aware of the possibility of introducing features that are not present in the original data; such artifacts are considered in the image enhancement case study in Chapter 11.

If filtering is used to make cosmetic improvements in images for publication, then report this.

## 4.5 Chapter Summary

In this chapter, the frequency domain approach to image filtering was introduced. The contributions of high and low spatial frequencies to the appearance of an image were discussed and this led to a generalized view of filtering in the frequency domain. Implementation of the generalized view using the Fourier transform was described, and the reader performed frequency domain filtering using ImageJ. The relationship between convolution filtering and filtering in the frequency domain was discussed and investigated in an activity using ImageJ.

## 4.6 Feedback on the Self-Assessment Questions

### Self-assessment question 4.01

The image shows the femur, with the femoral head and neck. The pelvis is in the top right of the picture. The image was acquired using the dual energy x-ray absorptiometry (DXA) technique, which uses x-rays of two distinct energies to generate images of bone and soft tissue. Measurements from DXA image data are used to determine bone mineral density (BMD) in the investigation of osteoporosis.

### Self-assessment question 4.02

The order should be (b) then (a) then (c). The greater the number of bands there are in a fixed distance, then the higher the spatial frequency.

### Self-assessment question 4.03

The image in Figure 4.3b has fewer high spatial frequencies, as there are fewer details seen in the image.

### Self-assessment question 4.04

A high-pass filter allows high spatial frequencies through and blocks the low spatial frequencies. High spatial frequencies carry information about small details in the image. Only (b) has small details and noise remaining, thus this is the image that has had the high-pass filter applied.

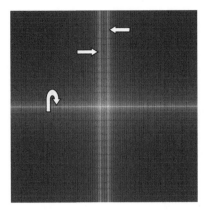

**FIGURE 4.24**
Annotated spectrum for feedback on self-assessment question 4.05. The curved arrow indicates the line representing zero frequency in the y direction, the straight arrows show the features representing the dominant x frequency.

### Self-assessment question 4.05

The line representing the frequency of zero in the y direction is indicated using a curved arrow in Figure 4.24. The two bands representing the x frequency are shown with straight arrows. It is emphasized again that the spectrum includes both positive and negative frequencies for both x and y, and the one dominant frequency in the x direction is plotted for both positive and negative values. The origin of the graph (0,0) is at the center.

### Self-assessment question 4.06

After editing, the parts of the spectrum representing the stripes have been retained, as has the zero-frequency y component. The only things removed are higher frequency components in the x direction. It is expected that in the result there will be stripes at the same frequency as in the original image, with a little less noise. The result is shown in Figure 4.25b, with the original shown in Figure 4.25a for comparison. This example illustrates how much information was held in the relatively small proportion of the spectrum that was retained. As predicted, we have the original stripes, with slightly less noise. It is useful to be able to think about filtering in this way as it can help in both planning an approach and for checking that an operation has performed as expected.

### Self-assessment question 4.07

Filter (b) goes with image (d) and filter (c) with image (e). The filter (b) is the high-pass filter as the white areas (which will remain unchanged) correspond with the high frequency parts of the spectrum.

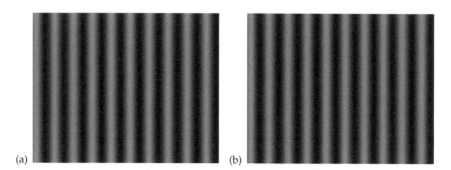

**FIGURE 4.25**
Images for feedback on self-assessment question 4.06. (a) Original image. (b) Result of frequency domain filtering using the filter shown in Figure 4.8.

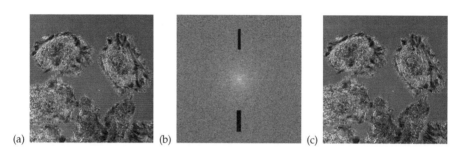

**FIGURE 4.26**
Images for feedback on self-assessment question 4.08. (a) Original image with stripes, (b) edited spectrum, (c) result of frequency domain filtering. (From http://rsb.info.nih.gov/ij/docs/examples/FFT/. With permission).

### Self-assessment question 4.08

The original striped image is shown in Figure 4.26a. The edited spectrum, with two important areas removed by using the Rectangle selection tool and Edit | Clear, is in Figure 4.26b. The result of this fairly rough editing is in Figure 4.26c.

### Self-assessment question 4.09

The only true statement is c). The FFT of the image is multiplied by the frequency domain filter.

### Self-assessment question 4.10

All of the five statements are true. You may have thought that b) and e) about enhancement were false because the emphasis in this chapter has been on the removal of features, but the spectrum can also be adapted to enhance features.

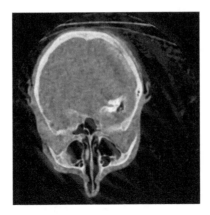

**FIGURE 4.27**
Filtered image for feedback on self-assessment question 4.11. The unfiltered image appears in Figure 4.23. (Image data from the Chapel Hill Volume Rendering Test Data Set, Volume I. [1] With permission.)

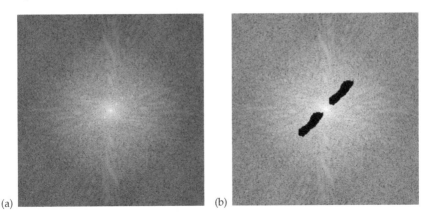

(a)   (b)

**FIGURE 4.28**
Images for feedback on self-assessment question 4.11. (a) The spectrum of the striped image of Figure 4.23 before editing, and (b) after editing. (Image data from the Chapel Hill Volume Rendering Test Data Set, Volume I. [1] With permission.)

## Self-assessment question 4.11

An example of the result obtained is in Figure 4.27. To obtain this result, the following steps were taken: The image was opened using ImageJ, and a Fast Fourier Transform was applied using the Process menu (Figure 4.28a). The series of bright dots running diagonally across the center of the image, from bottom left to top right, represent the stripes in the original image. The FFT was edited to remove the bright dots by selecting the freehand selections tool and drawing on the image to outline the dots in each quadrant, then using Edit | Clear to set the inside to zero (Figure 4.28b). An inverse FFT was performed on the edited spectrum using Process | FFT | Inverse FFT. Note how the direction of the stripes in the image was different from the example confocal microscope image considered in a previous question. This meant that the features in the FFT were also different. Although, with thought, one can predict roughly where the features will appear, it may sometimes be necessary to use trial and error to identify exactly which features in the spectrum are responsible for features in the images.

## Reference

1. The Chapel Hill Volume Rendering Test Data Set, Volume I, Department of Computer Science, University of North Carolina. Available at http:// education.siggraph.org/resources/cgsource/instructional-materials/volume-visualization-data-sets

# 5

## *Image Analysis Operations*

### 5.1 Introduction

The image analysis operations discussed in this chapter are the essential operations that are used in most quantitative image analysis protocols, including the basics such as addition and subtraction. The operations are often used in protocols designed to perform a particular calculation, and in this case a series of actions may be performed to accomplish the goal. Alternatively, an operation is sometimes used in isolation, for example, to generate a test image with preplanned gray values, or to normalize or calibrate the gray levels in an image. Image arithmetic operations are covered first, followed by logical and morphological operations applied to binary images.

#### 5.1.1 Learning Objectives

When you have completed this chapter, you should be able to

- Understand that arithmetical operations may involve one image and a constant, or be performed using pixel values in two images
- List several commonly used image arithmetic operations
- Recognize the most appropriate image data format for the result of an image arithmetic operation
- Define the logical operations that may be performed on binary images
- Explain the role of the structuring element in morphological operations
- Use ImageJ to perform a range of arithmetic, logical and morphological operations.

#### 5.1.2 Example of an Image Used in This Chapter

*Self-assessment question 5.01*

Figure 5.1 shows an image that is used later in the chapter. Which imaging modality has been used to acquire this image? What is the name of the unit associated with this imaging modality, which is defined to have a value of zero for water?

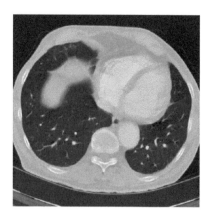

**FIGURE 5.1**
Image for self-assessment question 5.01.

## 5.2   Image Arithmetic

Familiar arithmetical operations, such as addition, subtraction and multiplication, can be performed on the pixel values of digital images. The approaches may conveniently be grouped into two categories:

- Operations that involve one image, and where appropriate, a numerical constant. For example, the constant value 15 might be subtracted from every pixel value.
- Operations that involve two images. Operations are performed using the pixel values at corresponding locations in each image. For example, the pixel value at point A in one image might be added to the pixel value at point A in the other image, with the operation repeated for every pixel in the image.

In ImageJ, the first set of operations appears in the Process | Math menu and the second in the Process | Image Calculator menu.

### 5.2.1   Operations on Data in One Image

#### 5.2.1.1   *Typical Operations*

Common operations that involve one image and a numerical constant are add, subtract, multiply and divide. Figure 5.2a shows the original image, and Figure 5.2b–e the results for a range of basic operations. Any operation that can be performed on a number is possible. Those commonly available in image processing packages include minimum, maximum, logarithms

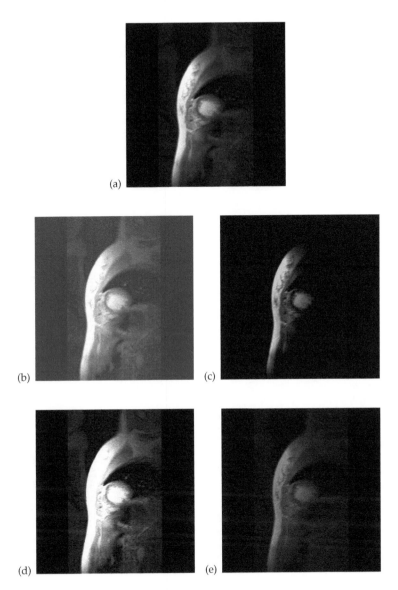

**FIGURE 5.2**
Arithmetical image operations. (a) Original, (b) addition of 64, (c) subtraction of 64, (d) multi-plication by 2, (e) division by 2.

(natural and base10), exponential, reciprocal, absolute value, square and square root. Four examples are shown in Figure 5.3. The minimum and maximum operations do not simply return an image showing the minimum or maximum pixel value in the original image. Instead, a numerical constant is required and the result is an image where pixels with a value less than, or greater than, the specified constant are replaced by the constant.

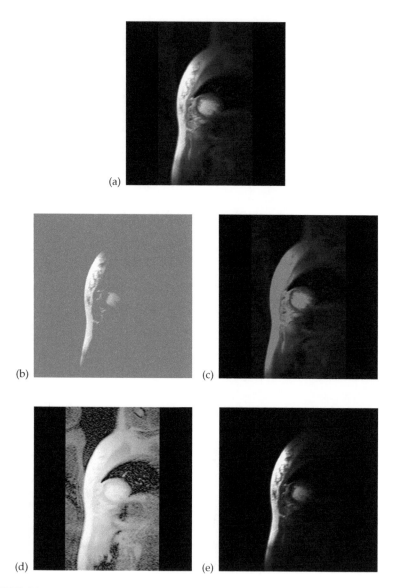

**FIGURE 5.3**
More arithmetical image operations. (a) Original, (b) minimum compared with 128, (c) maximum compared with 128, (d) natural logarithm, (e) square.

### 5.2.1.2   Data Type

The data type is important because of the potential for the calculated value to lie outside the range of the original image data type. For 8-bit images, the maximum allowable pixel value is 255. Values greater than 255 are usually set to 255 when the operation is performed, and this was the case for the

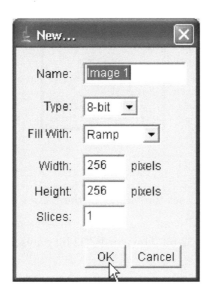

**FIGURE 5.4**
The ImageJ New Image dialog.

images in Figure 5.2. Similarly, unsigned data types do not allow values less than zero, although negative values can easily arise when subtraction is performed. If an unsigned data type is used, values less than zero will be set to zero. This effect is called *saturation,* where out-of-range pixels are set to the extremes of the allowable range.

### Activity: Being Prepared for Negative Numbers when Performing Image Arithmetic in ImageJ

Start ImageJ.

Select File | New and set the values in the New ... dialog box to those shown in Figure 5.4. Click on OK. The image size and data type are displayed immediately above the image.

Run the cursor across the image from left to right, watching the display of pixel values shown at the bottom left of the ImageJ menu. You will see that the values range from 0 to 255.

Select Process | Math | Subtract. Enter 128 in the Subtract dialog box (Figure 5.5) and click on OK.

Run the cursor across the resulting image, and you will see that the value is 0 until halfway across and the maximum is 127. This is consistent with results less than zero being set to zero—in other words, saturation has occurred.

Close Image 1 without saving.

**FIGURE 5.5**
The ImageJ Subtract dialog.

Select File | New and set the values in the New ... dialog box exactly as before and click on OK.

Select Image | Type | 32-bit. The words "32-bit grayscale" now appear above the image.

Perform the same subtraction operation as before. Run the cursor across the result. This time the values range from −128.0 to 127.0. By setting the data type to 32-bit, which is a data type that supports negative numbers, saturation has been avoided.

Close Image 1 without saving and exit from ImageJ.

*Self-assessment question 5.02*

Figure 5.6a shows the pixel values in a 3×3 pixel area of an unsigned 8-bit image. Figures 5.6b–e show the results of the four different operations listed below on the image. Match each of the following four operations to the correct figure letter.

- Image − 50
- Abs (Image)
- Image × 2
- Image + 255

*Self-assessment question 5.03*

Figure 5.7a and b show the pixel values in a 3×3 pixel area of two unsigned 8-bit images. If each image is multiplied by 4, which image will have some saturated pixel values if the output image data type is unsigned 8-bit?

**Activity: Using the ImageJ Math Menu**

The image syntheticflow.tif* is on the CD; open the image in ImageJ. This is a synthetic transverse MRI flow image of the neck, prepared for this activity.

---

\* Image data courtesy of John Ridgway. With permission.

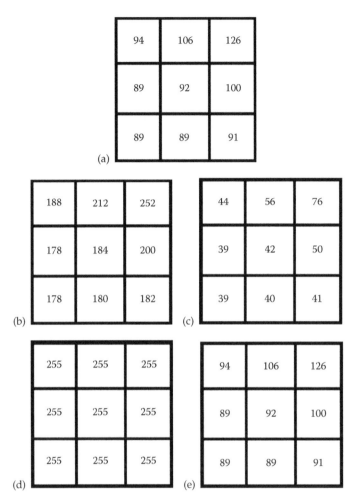

**FIGURE 5.6**
Images for self-assessment question 5.02. (a) Original pixel values in a 3×3 pixel area of an unsigned 8-bit image, (b)–(e) results of different operations.

| 134 | 154 | 169 |
|-----|-----|-----|
| 106 | 126 | 148 |
| 92 | 63 | 0 |

(a)

| 40 | 44 | 0 |
|----|----|----|
| 56 | 42 | 50 |
| 39 | 63 | 41 |

(b)

**FIGURE 5.7**
(a)–(b) Images for self-assessment question 5.03.

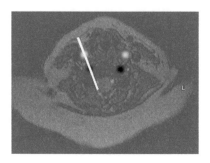

**FIGURE 5.8**
A synthetic transverse MRI flow image of the neck, for the activity on using the ImageJ Math menu. The flow information is artificial and differs from the in vivo case where the flow in all the blood vessels is in the same direction. (Courtesy of John Ridgway. With permission.)

The pixel values in the blood vessels (the four circular areas) have been added artificially and do not represent the true situation; in real life the flow in all four vessels is in the same direction. Run the cursor over the image, paying attention to the four circular areas of pixels with more extreme values. You should notice that the values are negative in the lower pair and positive in the upper two. This is the simulation of blood flowing in opposite directions.

Open two further copies of the image. On one, select Process | Math | Abs. On the other, select Image | Type | 8-bit. Both these operations give an image without negative pixel values.

For each of the three images, select the straight line selections tool, and place a line across the image so it crosses two of the vessels where the flow is in opposite directions (Figure 5.8) and then select Analyze | Plot profile.

Note that although the directional information has been lost when the Abs option is used, the magnitude of the pixel values is retained. It is possible to assess the relative size of the flow in each vessel. In the image that was simply converted to an unsigned 8-bit image, it is not possible to identify what pixel value corresponds with zero flow, though it is clear that it is likely that the flow is in opposite directions. If it is necessary during image analysis to convert signed data to unsigned data, it is worth thinking ahead about how the data are to be used in later calculations. Ensure that information is not inadvertently discarded.

### 5.2.2   Operations Using Data from Two Images

#### *5.2.2.1   Typical Operations*

All of the operations listed previously involving a numerical constant and one image can also be performed using the pixel value from the same location in a second image instead of a constant. Additionally, with a pair of

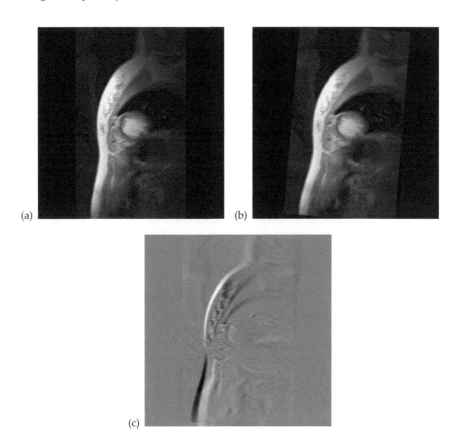

**FIGURE 5.9**
Arithmetical image operations involving two images. (a) First image, (b) second image, (c) result of subtraction. In midgray areas the result of subtraction is zero. Black and white represent large negative and large positive results.

images there is the option to perform more complex calculations, for example, finding the average of two images, or a weighted sum.

Figure 5.9 shows the result of subtracting a pair of slightly misaligned images. A similar example, the result of subtracting a pair of binary images, has previously been presented in Figure 2.22.

### 5.2.2.2 Data Type

The same considerations about data type apply to two image operations as they do to operations between one image and a constant. Again, it is necessary to convert to a signed data type, or one with a larger dynamic range, to avoid saturation.

### Activity: The ImageJ Image Calculator

In this activity, ImageJ is first used to generate reference images with known pixel values. These images are then used in the Image Calculator.

Start ImageJ. Select File | New | Image. In the dialog window, set Type to 8-bit, Fill With to Ramp, Width to 256 pixels, Height to 128 pixels and Slices to 1. Click on OK.

Select the new image, then select Image | Adjust | Canvas Size. In the dialog window set both Width and Height to 256 pixels, Position to Center and tick Zero fill. Click on OK.

Ensure that the ImageJ menu box is visible, then run the cursor over the newly created image, noting how the pixel value is zero in the top and bottom bands, and runs from 0 to 255 in the ramp area.

Select Image | Duplicate to make a second copy of the ramp image, then select Image | Rotate | Rotate 90 degrees Left. There are now two images containing a range of gray values in a well-understood configuration, which may be used with the ImageJ Image Calculator.

Save both images in the .tif format with explanatory file names. Close the open images, and load the two recently saved files.

Select Process | Image Calculator. In the dialog box, set Image1 to the name of the image with the horizontal ramp, and Image2 to the name of the other ramp image. Set the Operation to Add, tick beside Create New Window, but do not tick next to 32-bit result. Click on OK.

Repeat the previous step, but this time place a tick next to 32-bit result. The results should appear as is shown in Figure 5.10.

Run the cursor over each image, noting the differences especially in the upper right part of the central square. Think about why this area is different in the two images.

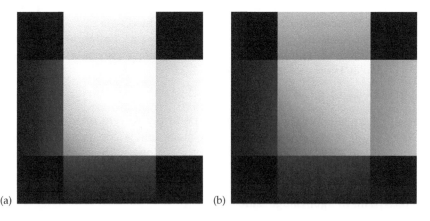

(a)                                      (b)

**FIGURE 5.10**
Addition in the ImageJ calculator. (a) 8-bit result, (b) 32-bit result.

Generate two more images in a similar way, this time using the subtract operation. Compare the images.

*Self-assessment question 5.04*

For the pair of images generated and saved in the previous activity, predict the pixel value expected for the pixel at x = 127, y = 128 (which has the value 127 in both images) when the multiply operation is used, both without and with the 32-bit result box ticked. Check the answer using ImageJ.

*Self-assessment question 5.05*

The Image Calculator in ImageJ offers both a subtract and a difference operation. Find out how these differ.

### Investigating the Image Calculator

Use the two reference images to produce a full set of results from the Image Calculator, performing each of the operations with both 8-bit and 32-bit results.

## 5.2.3   Spatial and Gray Level Calibration

Both spatial calibration and gray level calibration are implemented using the arithmetical operations described in this chapter. It is assumed that the image being calibrated is free from geometrical distortions and gray level inhomogeneity. The effects of these factors are discussed in Chapter 7.

### 5.2.3.1   Spatial Calibration

Spatial calibration ensures that the real-life size is associated with each pixel of an image. For example, if a millimeter scale is imaged alongside the object, it is possible to determine the distance on the scale occupied by a known number of pixels. Calibration is achieved by dividing the known distance by the number of pixels to give the size of each pixel in the object space. Errors are reduced by using as large a number of pixels as is practical.

   The calibration procedure is automated in ImageJ. The user first selects the straight line selection tool to draw a line on the image that corresponds with a known distance. Then the Analyze | Set Scale dialog is used to enter the known distance and its unit of measurement. The distance in pixels will have been set automatically. Alternatively, if pixel size information has been supplied with an image, there is no need to define a distance using the straight line tool. In the Analyze | Set Scale dialog, enter 1 for the distance in pixels, and enter the pixel size as the known distance together with its unit of measurement. Images imported in the DICOM format (Chapter 6), do not need calibration, as the pixel size is one of the pieces of information stored with the image data.

*Self-assessment question 5.06*

You are told that each pixel in chapter5.bmp (which is on the CD) measures 0.58 mm × 0.58 mm. Load and spatially calibrate chapter5.bmp in ImageJ. What is the width of the image in millimeters?

### 5.2.3.2  Gray Level Calibration

In an uncalibrated image there is no guarantee that, in a pair of images from the same modality but acquired at different times, the same gray value is associated with the same object material. Or alternatively, that the same object material appears with the same gray value in all images. Gray-level calibration ensures that it is known what value of real-life object property (such as attenuation coefficient) is represented by a pixel value. In spatial calibration, a millimeter scale may be imaged to provide known distances, and for gray level calibration the equivalent is to image a reference standard. Reference standards include a step wedge for radiographs and radioactive standards for radionuclide imaging. It is then possible to plot the relationship between the pixel value and the known standard values, either manually or within a program such as ImageJ (using Analyze | Calibrate). Although ideally several reference standards spanning the whole dynamic range would be used, if the gray scale is assumed to be linear, just one reference plus the background may be applied.

Image normalization is a less rigorous variation that is appropriate if images from the same individual are to be compared with each other. An imaged object that is not expected to change between imaging sessions is required in all the images. Each image is normalized by dividing all the pixel values by the gray value of the unchanging object in that particular image. The choice of the unchanging area of the image requires some thought. For example, although a metal implant will not change its density between repeated x-rays, it is likely to have a saturated gray value in acquired images because it is very much denser than soft tissue. It would be better to choose an area of healthy bone for example.

---

## 5.3  Binary Image Operations

A binary image is one where just two gray levels are used. A binary image can be made up of any two gray values. For convenience in this discussion, the two values used are 0 and 1, where background pixels have the value 0 and object pixels are set to 1.

### 5.3.1  Logical Operations

Logical operations are performed using two binary images such as those in Figures 5.11a and b. At each pixel location, a comparison is made, which

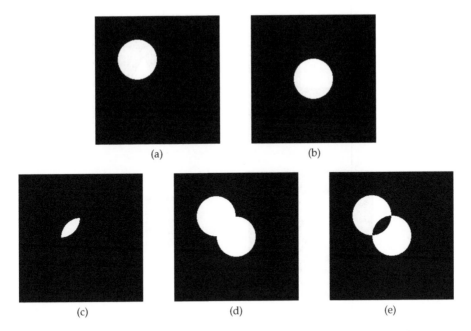

**FIGURE 5.11**
Logical operations with binary images. (a) First image, (b) second image, (c) result of AND, (d) result of OR, (e) result of XOR.

depends on the operation. The pixel value for the resulting image is set to 1 or 0 depending on the operator as follows:

For the logical AND operator: If the pixel is 1 in one image AND is 1 in the other image, then the result is set to 1; otherwise the result is set to 0 (Figure 5.11c).

For the logical OR operator: If the pixel is 1 in one image OR is 1 in the other image, then the result is set to 1; otherwise the result is set to 0 (Figure 5.11d).

For the logical XOR (eXclusive OR) operator: If the pixel is 1 in one image OR is 1 in the other image, but is not 1 in both, then the result is set to 1; otherwise the result is set to 0 (Figure 5.11e).

*Self-assessment question 5.07*

Which of the images in Figure 5.12c–g shows the result of the logical AND operation between the two binary images in Figure 5.12a and b?

*Self-assessment question 5.08*

Which of the images in Figure 5.12c–g shows the result of the logical OR operation between the two binary images in Figures 5.12a and b? Sketch the expected appearance of the logical XOR operation between the two binary images in Figures 5.12a and b.

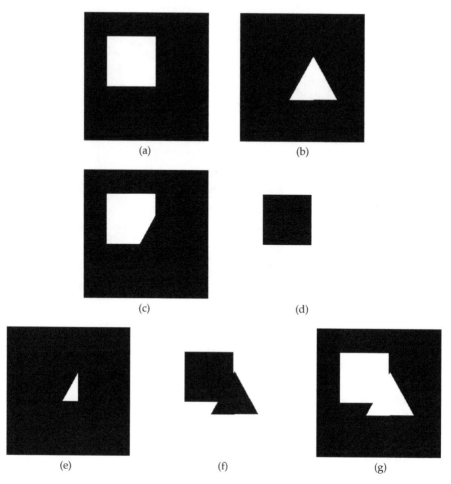

**FIGURE 5.12**

Images for self-assessment questions 5.07 and 5.08. (a) First image, (b) second image, (c)–(g) results of operations between the two images.

## Activity: Using Logical Operations in ImageJ

For this activity either use the images binary03.bmp and binary04.bmp supplied on the CD, or make your own images as indicated below.

### Custom Binary Images

To make your own binary images for the activity in ImageJ, use File | New | Image … and generate two 8-bit images that are the same size as each other and filled with black. Use one of the shape selection tools to draw a shape in one image, and fill it with white using Edit | Fill. Draw and fill a shape in the other image, ensuring that there is some overlap between the shapes in the two images. Save the two images and use them in place of binary03.bmp and binary04.bmp in the activity.

Load binary03.bmp and binary04.bmp from the CD into ImageJ. These images might be, respectively, a reference segmentation of an object in an image and a new segmentation of the same object. Logical operations can be used to compare the two.

Select Process | Image Calculator. In the dialog box, set Image1 to binary03.bmp, and Image2 to binary04.bmp. Set the Operation to AND, tick beside Create New Window, but do not tick next to 32-bit result. Click on OK.

The result has all the pixels that have been selected in both images: they could be termed the true positive pixels (Chapter 10) as they have been correctly selected in the new segmentation.

Repeat using the XOR operator. The resulting image shows the pixels that differ in the two images.

## 5.3.2 Morphological Operations

Morphological operations involve one image (most often a binary image) and a small binary template called a *structuring element*. The structuring element is positioned over every pixel in the image in turn. At each location a comparison is made between the structuring element and the pixels of the neighborhood under the structuring element, and a new image is generated that depends on the outcome of that comparison. The possibilities are that

- the structuring element *fits* the image (all of the underlying pixels have value 1)
- the structuring element *intersects* the image (one or more of the underlying pixels has value 1) or
- the structuring element does neither (none of the underlying pixels has value 1).

These possibilities are illustrated in Figure 5.13. It is the intersecting option that is used by the common morphological operators *dilate* and *erode*, which are described next.

For the dilate operator, the pixel in the new image will be set to 1 if the structuring element intersects the image. This means that background pixels neighboring an object are changed to object pixels in the result. Dilation is used to fill small holes and has the effect of enlarging objects (Figure 5.14b).

The opposite effect is achieved with the erode operator. In this case, the pixel in the result is set to 0 if the structuring element intersects the image. This means that all object pixels neighboring the background are changed to background pixels in the result. Isolated pixels or narrow regions are removed by erosion, and the size of objects is reduced (Figure 5.14c).

Note that the shape and size of the structuring element have an effect on the result, so it is always necessary to give information about the structuring

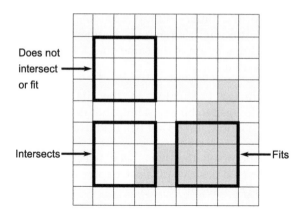

**FIGURE 5.13**
The terms associated with the use of the structuring element for morphological operations. A square 3×3 structuring element is shown in locations at which it fits, intersects and neither fits not intersects the binary object shown in gray.

element as well as the operation performed when recording or reporting such analysis.

The two other common morphological operators are *Close* and *Open*. Close is a Dilate followed by an Erode, while Open is an Erode followed by a Dilate. By using the operations in succession, the hole-filling or small feature removal can be achieved without a change in the overall size of the object (Figure 5.14d–e).

### Activity: Using Morphological Operations in ImageJ

The default setting for binary images in ImageJ (and many other packages) is black objects and white background. This is the opposite of the convention used in this chapter, where black has been used for background areas. It is important to set up any program properly, or a morphological operation may appear to have the opposite effect from that expected. In ImageJ the setting up is done using Process | Binary | Options. Ensure there is a tick next to Black Background. Note also that the structuring element used in ImageJ is a square 3×3 neighborhood (this information is given in the manual).

Load binary03.bmp from the CD into ImageJ.

Make a second copy of the image using Image | Duplicate.

Perform morphological erosion on the duplicate image by selecting Process | Binary | Erode.

Select Process | Image Calculator, and subtract the eroded image from the original image binary03.bmp. You will be left with an outline comprising the outermost pixels of the original binary object. The morphological operation has been used as a quick and simple edge detector.

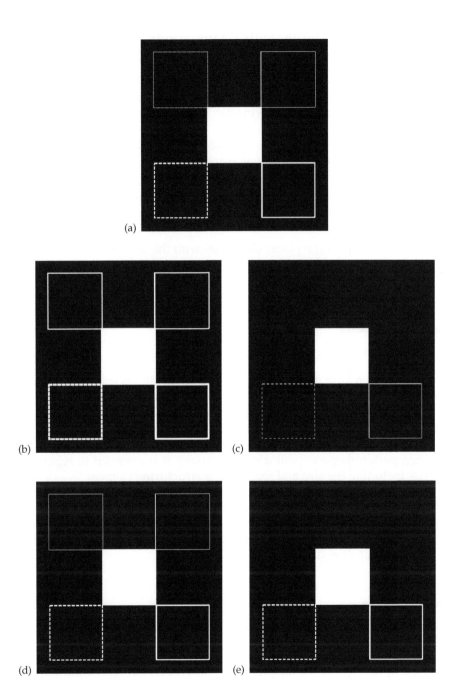

**FIGURE 5.14**
Morphological operations with binary images using a square 3×3 structuring element. (a) Original, (b) result of dilate, (c) result of erode, (d) result of close, (e) result of open.

## 5.4    Logical and Morphological Operations on Grayscale Images

The logical and morphological operations described above in relation to binary images may also be applied to grayscale images, but the results can be less easy to predict. The following can be expected:

- The logical operators AND, OR and XOR work by using the binary representation of the pixel value and performing the logical operation on each bit.

- Grayscale dilation replaces each pixel with the brightest value underlying the structuring element (assuming that a black background has been set).

- Grayscale erosion replaces each pixel with the darkest value underlying the structuring element (assuming that a black background has been set).

## 5.5    Good Practice Considerations

The analytical nature of the arithmetic operations discussed in this chapter means that they are less likely than the operations covered in other chapters to be applied without thought being given to the numerical values involved. As is always the case, however, any values chosen must be justifiable and be applicable to all images in a set being analyzed.

When performing morphological operations, it is essential to report the details, including the size and shape of the structuring element.

Just as one does when performing non-image arithmetic, it is wise always to think about the result expected and to check the result against these broad expectations (for example: should the result really be negative?). With images it may be sensible to check the operations on simple synthetic images before undertaking a full-scale analysis.

## 5.6    Chapter Summary

This chapter covered the key arithmetical, logical and morphological image analysis operations. It was emphasized that arithmetical and logical operations might involve one image and a numerical constant, or use the pixel values from two images. ImageJ activities were used to illustrate how the operations might be valuable in practical applications. Spatial and grayscale image calibrations were described.

## 5.7 Feedback on Self-Assessment Questions

**Self-assessment question 5.01**

The image was acquired using x-ray computed tomography. The Hounsfield unit (also known as the Hounsfield number, or CT number) is defined as −1000 for air and 0 for water.

**Self-assessment question 5.02**

(b) Image × 2, (c) Image − 50 (d) Image + 255 (e) Abs(Image)

**Self-assessment question 5.03**

Image (a) will have some saturated pixel values if the output image data type is unsigned 8-bit, because values greater than 255 will be present. In image (b), none of the pixels will give a value over 255 when multiplied by 4.

**Self-assessment question 5.04**

Without 32-bit data type selected, the pixel value expected is 255. This is an image that will display saturation. With 32-bit data type selected, the pixel value expected is 16129.

**Self-assessment question 5.05**

The information may be found in the ImageJ manual. The Difference result is the absolute value of the result of subtraction.

**Self-assessment question 5.06**

Use Analyze | Set Scale … The width of the image is 296.96 mm. After spatial calibration, this can either be read from the top of the image or measured using the straight line selection tool.

**Self-assessment question 5.07**

Figure 5.12e shows the result of the logical AND operation between the two binary images in Figure 5.12a and b.

**Self-assessment question 5.08**

Figure 5.12g shows the result of the logical OR operation between the two binary images in Figure 5.12a and b. The XOR operation will result in an image similar to Figure 5.12g, but the area where the two shapes overlap will be black.

# 6

## Image Data Formats and Image Compression

### 6.1 Introduction

Image data formats are an essential part of image processing activity because they define how the image data are stored and which additional information is saved together with the image data. Image compression is often a component feature of an image file format because it can reduce the size of the file for storage and transmission. An understanding of different ways of achieving compression is valuable to avoid the possibility of inadvertently changing pixel values through an inappropriate choice of method.

#### 6.1.1 Learning Objectives

When you have completed this chapter, you should be able to

- Define the term image format and explain why general purpose image formats are unsuitable for medical imaging
- Be aware of the medical imaging standard DICOM®*
- Explain why data compression is often necessary for image files
- Describe the general principle underlying image compression
- Distinguish between lossy and lossless compression techniques
- Outline the run length encoding, Huffman and jpeg methods
- Use ImageJ to import and save images in various formats

#### 6.1.2 Example of an Image Used in This Chapter

*Self-assessment question 6.01*

Figure 6.1 is a digital mammogram from the Digital Database for Screening Mammography [1,2]. The images stored in the database were digitized from film and have a pixel size of around 50 μm. If a digitized mammogram

---

* DICOM is the registered trademark of the National Electrical Manufacturers Association for its standards publications relating to digital communications of medical information.

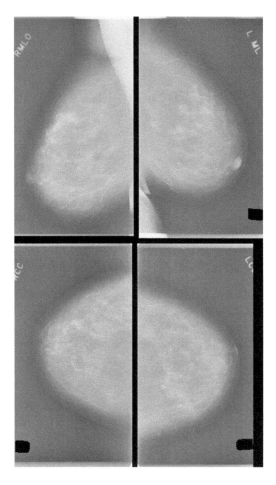

**FIGURE 6.1**
Example of a digital mammogram. (From the Digital Database for Screening Mammography [1,2]. With permission).

measured 4000×5000 pixels, and each pixel was stored as a 16-bit number, how large would the file size be?

## 6.2 Image Data Formats

### 6.2.1 The Need for Image Data Formats

A digital image is a computer file containing a list of numbers that represent the gray values of the pixels in an image. If you think about this statement, a number of questions arise about the correct interpretation of this file full of numbers. How many pixels are there in each row of the image? How many rows? Is the first number in the file from the top left of the image or somewhere

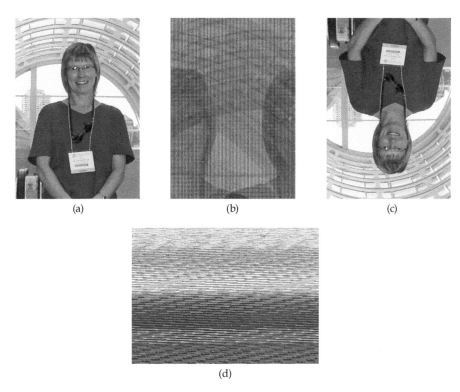

(a)  (b)  (c)

(d)

**FIGURE 6.2**
Examples of the effect of wrong assumptions about the format of an image. (a) Original image, (b) 16-bit data read as 8-bit data, (c) data placed from bottom to top instead of from top to bottom, (d) image width and height exchanged. Image formats define standardized locations in the header of the image for important information like this.

else? Is each pixel value represented by 1 byte of data, as for the 8-bit data type? Or is this a 16-bit image where every pixel is represented by 2 bytes? Or is it a color image? Is there more than one image in the file? The consequences of answering some of these questions wrongly are illustrated in Figure 6.2.

Image data formats are a way to avoid mistakes in interpretation arising from such uncertainties. Formats describe a standard way in which image data are stored, so that an image file may be correctly interpreted. The existence of an image data format means that different systems can read and write the same image files, and image storage and interchange are facilitated

Generally an image file has two parts, the image file header and the image data.

## 6.2.2 Image Header

The header contains the information necessary to allow the original image data to be reconstructed from the pixel values stored in the second part of the file:

(a)                              (b)                              (c)

**FIGURE 6.3**
The introduction of a left-right flip through placement of the data. (a) Original image, in a format where the first pixel is the one at the bottom left, (b) the data loaded with the first pixel at the top left, (c) the preceding image rotated through 180°.

- Number of pixels wide
- Number of pixels high
- Data type (gray scale, color)
- Number of bits per pixel

The pixel order is not usually stated in the header, but is consistent for a particular image format. Most formats order the pixels from left to right across rows and start at the top left. However, some formats are ordered starting from the bottom left. If such an image is loaded assuming that the first pixel is at the top left, a result like that in Figure 6.3b is obtained. The result may appear simply to be rotated through 180°, but applying such a rotation leads to the image in Figure 6.3c. In fact, the image is flipped from left to right compared with the original: look at the sunglasses and the asymmetrical roof. Clearly the wrong choice of starting pixel can have significant effects. The result is equivalent to a flip about the y axis, and if performed on clinical transverse slices would mean that left-right reversal occurs. An individual's heart could appear to be on the wrong side of their body.

In the next two activities, two very simple image data formats are considered, text and raw data. Text data could, for example, be a 2-D array of numbers saved from a spreadsheet. Raw image data will often come from another application. To import such data correctly, the size of the image and the size of any header must be known together with the type of image data (e.g., 8-bit, 16-bit).

## Activity: Importing an Image from a Text File into ImageJ

Start ImageJ.

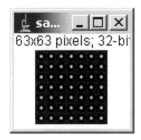

**FIGURE 6.4**
Sample_text_image.txt when imported as a text image into ImageJ.

Select File | Import |Text Image to import an image from a text file. Choose the file sample_text_image.txt from the CD, and click the Open button (Figure 6.4).

### A Closer Look at the Text Image

Use a text viewer (Microsoft Notepad®* or similar) to view the contents of the sample_text_image.txt file. The values in the file are separated by tabs (this is known as a tab delimited file), and there are the same number of lines of data in the file as there are rows in the image. The text image import operation is described in the File Menu section of the ImageJ Documentation. There are many more options for importing images; see the File Menu section of the documentation for details. Note that although ImageJ can import TIF files, they must be in an uncompressed version of the format.

## Activity: Importing an Image from a Raw Data File into ImageJ

Close any open images in ImageJ or start the program.

Select File | Import | Raw. Choose the example file, which is called sample_raw_image.pgm, from the CD.

A dialog box entitled Import ... will appear. The image is known to be 288×324 in size, is stored as 8-bit data and has a header of 46 bytes (thus the image data are "offset" by this many bytes from the top of the file). Enter the correct values for width, height and offset into the dialog box, leaving the number of images at 1 and the gap at 0. Do not tick the three tickboxes. Click on the OK button and check that the image looks like Figure 6.5.

Select File | Quit to close ImageJ. You will be asked if you wish to save any open images.

---

* Registered trademark of Microsoft Corporation, Redmond, Washington, USA.

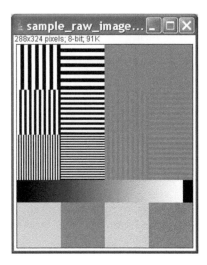

**FIGURE 6.5**
Sample_raw_image.pgm when imported as a raw image into ImageJ.

### 6.2.3   General Purpose Image Data Formats

You may already be familiar with general purpose image formats such as bmp, wmf, tif, gif, jpg. These can be very useful for saving images for use in presentations and documents, but they are not suited for primary storage of patient data, or for image data that are to undergo image analysis. The reasons they are unsuitable are as follows:

- Some formats incorporate lossy compression, which means that pixel values are changed
- Many formats cannot save 16-bit grayscale images which are common in medical imaging
- The file headers do not have a means to save patient and examination details
- Few formats are designed for multislice data sets
- Basic information such as pixel size is not saved

### 6.2.4   DICOM®

The limitations of general formats mean that medical imaging systems use special formats that allow space to store not just image data but also the associated metadata about patient, equipment and acquisition. Most manufacturers have proprietary, internal formats for storing their data, but also offer the option of the DICOM standard. The acronym DICOM stands for Digital Imaging and Communications in Medicine, and all medical imaging equipment purchased today should be DICOM compliant. DICOM involves more than just standards for stored data, but also includes communication standards.

### Find Out More about DICOM

The DICOM® format is very important, but the information in this chapter is necessarily a very basic introduction. The following resources may be consulted for more information.

- http://medical.nema.org/dicom.html
- http://www.dclunie.com/
- http://www.rsna.org/Technology/DICOM/index.cfm

Note that DICOM is often referred to as DICOM 3.0. The preceding two versions were called ACR-NEMA, so there was never a DICOM 1.0 or 2.0.

## Activity: The Wealth of Information in a DICOM File

In this activity, a publicly available dataset* in the DICOM format is opened in ImageJ.

Use File | Open in ImageJ and navigate to the Chapter06 folder on the CD. Select the COMUNIX folder, and continue to navigate down through the folders until you reach a folder called CT HEAD-NK 5.0 B30s, which contains files with the .dcm extension. Select any .dcm file in this folder and open it.

Select Image | Show Info … A window will open with the name Info for X.dcm (where X is the name of the particular file you chose to open). Scroll through this information and note the level of detail stored.

Repeat for an image from the PET WB folder, noting how the information stored differs for an image from a different imaging modality.

Click on one of the images to select it. Select File | Save As. Choose the bmp format and save the image.

Close the open windows in ImageJ, then use File | Open to open the bmp format version that you just saved.

Select Image | Show Info … The wealth of information in the DICOM header has been lost, and if the image was initially in 16-bit format it is now an 8-bit image.

When sharing medical images in clinical practice or research (within the restrictions of ethical approval), DICOM is much the best method. Even with DICOM or other known image formats, it is wise to exchange as much supporting information as possible when transferring image data, especially regarding compression, which must never be lossy. Ask for hardcopies of the original images to check the orientation and aspect ratio in your results.

---

* Image data from http://pubimage.hcuge.ch:8080/. With permission.

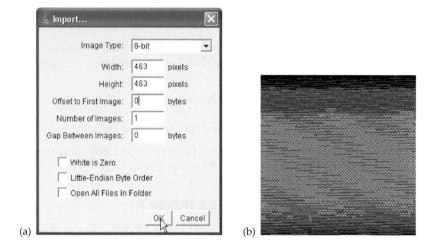

(a)                                                    (b)

**FIGURE 6.6**
(a) Values for importing square 8-bit image. (b) Image imported using the values shown in (a).

### 6.2.5 The Effect of Missing Information

In the following activity, some of the consequences of not having the necessary information about image size and data type are explored. The data file on the CD (unknownimage.raw) contains 215,168 bytes of data.

### Activity: Importing an Image with Unknown Format Using ImageJ

Select File | Import | Raw and select unknownimage.raw. Complete the dialog box as shown in Figure 6.6a, which contains values associated with the image data being unsigned 8-bit with no header and the image square. The result (Figure 6.6b) has a striped appearance that is characteristic of 16-bit data being read as 8-bit, and the rather unstructured pattern suggests that the size of the image is not correct either.

Select File | Import | Raw and select unknownimage.raw. Complete the dialog box as shown in Figure 6.7a, which contains values associated with the data being unsigned 16-bit with no header and the image square. The result (Figure 6.7b) looks very promising. You could go on to investigate the effect of selecting the little-endian byte order option and the signed 16-bit image type, especially if it is expected that the data have negative values (e.g., flow data).

It is instructive to see the effect of using close, but wrong, assumptions. Try importing at 8-bits with width 328 and height 656 (Figure 6.8a). The striped appearance is characteristic of 16-bit data being read as 8-bit, but with the correct image width. Now use 16-bit unsigned with width 656 and height 164 (Figure 6.8b). This image appearance arises because each row in the result is made up by concatenating two rows of an image that is really half the width. The images appear squashed from top to bottom as each is made up of only

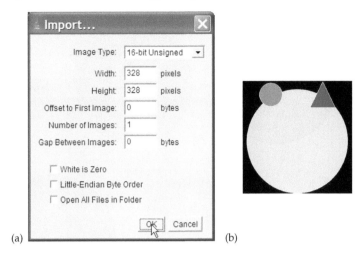

(a)                                          (b)

**FIGURE 6.7**
(a) Values for importing square 16-bit image. (b) Image imported using the values shown in (a).

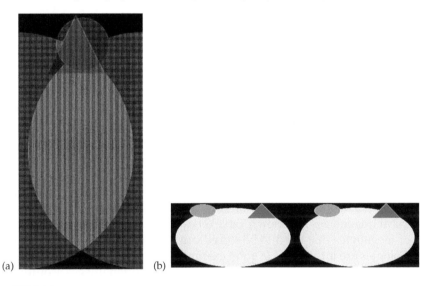

(a)                      (b)

**FIGURE 6.8**
(a) 16-bit image imported as 8-bit, but with correct image width. (b) 16-bit image imported with double the correct width, but with correct data type.

half the rows it should be. A result like this suggests that the data type is correct, but that the image dimensions are wrong.

*Self-assessment question 6.02*

The original image shown in Figure 6.9a is stored in a format where the first pixel is at the top left of the image. If the stored image is read by a system where the first pixel is at the bottom left of the image, which of the images b–e will be seen?

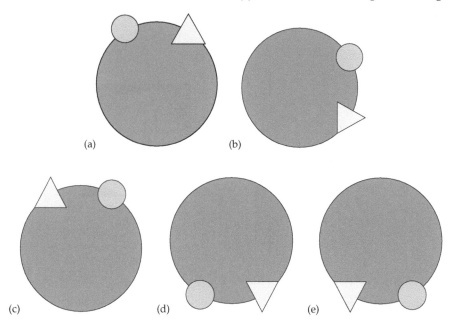

**FIGURE 6.9**
(a)–(e) Images for self-assessment question 6.02.

## 6.3   Image Compression

Image compression techniques reduce the amount of data needed to represent an image. This reduction is necessary because it helps make storage and transmission of data more efficient, for example, for storage in Picture Archiving and Communication Systems (PACS), for transmission in teleradiology and telemedicine applications and for sending images by email.

Typical image sizes are shown Table 6.1, where it can be seen that the number of megabytes that need to be stored can be very large. As new imaging technologies are introduced (a good example is multislice CT), the amount of data per examination increases. The very high-resolution images acquired in digital mammography are each very large even though few are acquired.

**TABLE 6.1**

Typical Image Sizes for Common Medical Imaging Modalities

| Modality | Image size | Bits per pixel | Images per examination | Mbytes per examination |
|---|---|---|---|---|
| CT | 512 × 512 | 12 | 100 | 40 |
| MRI | 256 × 256 | 12 | 100 | 10 |
| CR | 2048 × 2048 | 12 | 4 | 25 |
| SPECT | 128 × 128 | 16 | 50 | 2 |
| Ultrasound | 512 × 512 | 8 | 50 | 13 |

### 6.3.1 Compression Ratio

Image compression is based on the idea of redundancy. This is when more data are stored than are strictly necessary to convey the information:

- If an image contains many instances of the same pixel value, it is redundant to list that value over and over again, if its presence can be recorded in a more compact way.
- If an image contains values that could be stored using fewer bits than in a standard 8- or 16-bit format, then there is redundancy in the coding.
- If an image contains detail at a higher level of resolution (spatial or gray level) than can be perceived by the eye, then this detail may be considered redundant if the image is only to be viewed.

When image compression is used, the size (in bytes) of the image file is reduced. The *compression ratio* is defined as follows:

Compression Ratio = Original Image Size/Compressed Image Size

The larger the compression ratio, then the greater the amount of compression, and the smaller the compressed file size. If compression has successfully reduced the file size, then the compression ratio should be greater than 1.

The following example illustrates the general principles behind compression techniques. Consider an image of 640×480 pixels and comprising 32 black stripes of value 0 and 32 white stripes of value 255 (Figure 6.10). Stored as an 8-bit image, the image data require 640 × 480 × 8 bits = 307,200 bytes. However, there is clearly a lot of redundancy in this image, with the black and white values being repeated over and over again. In order to compress the data, a way of expressing the same information but using fewer bytes is needed. One solution would simply be to describe the image in words: "32 vertical white lines interleaved with 32 vertical black lines, each line

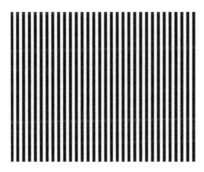

**FIGURE 6.10**
Image to illustrate the principles of image compression. It consists of 640×480 pixels and has 32 vertical black stripes and 32 vertical white stripes.

having a width of 10 pixels and height of 640 pixels." This statement is made up of 128 characters. A character can be stored using a single byte, so the wordy description of the image could be stored using only 128 bytes. For this image, compression by description corresponds with a compression ratio of 307,200/128, that is, 2400:1, which is a very high compression ratio.

### 6.3.2 Lossy vs. Lossless Compression

There are two types of compression: *lossless* and *lossy*. In lossless compression, the image reconstructed from the compressed data is identical to the original image. Lossless compression is also called *reversible* compression. The use of lossless compression is clearly an advantage in many medical imaging applications, for example, where data are used for primary diagnosis or when the use of analytical techniques is planned.

In contrast, lossy compression results in a reconstructed image that differs from the original image, possibly only in a small way. Lossy compression is also called *irreversible* compression. This sort of compression can lead to much smaller file sizes than the lossless methods, which is why it is used in spite of its effect on the data. Images compressed using lossy compression might be suitable for certain applications, for example, for online teaching files where fast loading of the file over a network is more important than the reproduction of detail in the image.

*Self-assessment question 6.03*

Think of two applications, not already suggested in the text, which would require lossless compression, and two where lossy compression would be acceptable.

There are very many compression coding techniques available. To illustrate the basic idea, two lossless methods that are used for medical images are further discussed: Run length encoding and Huffman encoding. One lossy method, jpeg, is also covered.

### 6.3.3 Run Length Encoding (RLE)

The principle of run length encoding (RLE) is based on the observation that nearby pixels in an image tend to have the same gray value. Compression could be achieved by grouping *runs* of pixels with same gray value into a single *code*. For RLE, each single code has 2 bytes. One is the pixel value, and the other the number of pixels in the run. The conventional notation is to show this as: gray value | run length, for example, 255 | 2.

If for example, the following pixel values occur in an image file

$$191 \quad 191 \quad 191 \quad 192 \quad 192 \quad 0 \quad 0 \quad 0 \quad 0 \quad 0 \quad 0$$

In RLE coding, this would be replaced by: 191|3 192|2 0|6, which says that there are three 191s, then two 192s and 6 zeroes. Instead of the 11 bytes required for the original list of pixel values, we need only 6 bytes for storage. Note that:

- RLE can give compression ratios of about 1.5:1 on grayscale images
- It is a lossless method, as it will result in the reconstruction of exactly the same pixel values as in the original image.
- RLE reduces the *interpixel redundancy* in an image.

However, an effect known as *data explosion* can occur with RLE. This occurs in noisy image, where adjacent pixels have different values. If, for example, the gray value changes between every pixel in the image, each pixel must be a separate run, and be represented by 2 bytes of data. As each pixel was represented by 1 byte initially, the compressed image is twice the size of the original.

*Self-assessment question 6.04*

Express the following section of image data in run length encoding style. If both original and compressed versions are stored using 8-bit values, what is the compression ratio? Comment on the result.

| 255 | 255 | 254 | 255 | 200 | 200 | 150 | 150 | 150 | 150 | 150 | 100 |
|-----|-----|-----|-----|-----|-----|-----|-----|-----|-----|-----|-----|
| 101 | 150 | 150 | 180 | 179 | 180 | 180 | 180 | 180 | 180 | 175 | 170 |

## 6.3.4   Huffman Coding

The idea behind Huffman coding is to replace pixel values with a code that requires less storage space. For example, the code may be as small as 1 bit to replace an 8- or 16-bit pixel value. The shortest codes are given to the gray values that occur most frequently in the image, so the first step in Huffman coding is to calculate the image histogram and rank the frequency of occurrence of each gray level. Note that:

- Huffman coding can give compression ratios of about 1.5:1 to 2:1 on grayscale images.
- It is a lossless method, as it will result in the reconstruction of exactly the same pixel values.
- Huffman coding reduces the *coding redundancy* in an image.

## 6.3.5   JPEG Compression

JPEG, or jpeg, is an acronym for Joint Photographic Experts Group. The origin of the method is a clue that the compression algorithm may work better for photographs of general scenes than for medical images.

Jpeg is a lossy compression technique, which works on the idea of reducing *psychovisual redundancy* in an image. It works in a manner analogous to frequency domain low-pass filtering (Chapter 4), in that high-frequency information, which may represent detail that would not be perceived by the eye, is selectively removed. In jpeg compression, a transform is applied to the data and some of the high frequency components are discarded, thus reducing the amount of data that need be stored. To avoid blurring the whole image too much, the procedure is performed on small areas of the image at a time, in blocks of 8×8 pixels. By using small blocks, it is possible to make the process *adaptive* (Chapter 3). This means that the choice of components for disposal is dependent on the image data in that particular block. For example, if it is an area where the loss of high frequencies is expected to be noticeable, then they may be retained there.

Jpeg compression leads to characteristic artifacts in the reconstructed images:

- Blocking, because of the use of 8×8 pixel areas
- Blurring, because high-frequency components have been discarded
- Ringing at edges, because high-frequency components have been removed using a sharp cut-off frequency
- Changes to pixel values

The method works particularly badly in geometrical images or diagrams, which often contain isolated sharp boundaries. In the exaggerated example in Figure 6.11, it can be seen that the original and post compression images differ in blocks that straddle the sharp boundaries in the image. The high-frequency components that are necessary to reconstruct the boundaries correctly have been discarded, leading to blocking artifacts. In a more complex scene, the artifacts would be much less noticeable. These artifacts (sometimes known as jpeggies) are very recognizable. They occur in blocks because of the manner in which the algorithm is applied in blocks of 8×8 pixels, and affect details in the image.

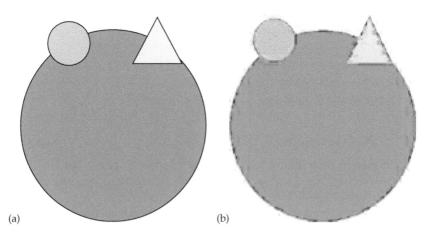

(a)                                    (b)

**FIGURE 6.11**
Demonstration of the effects of extreme jpeg compression. (a) Original image, (b) post-compression image.

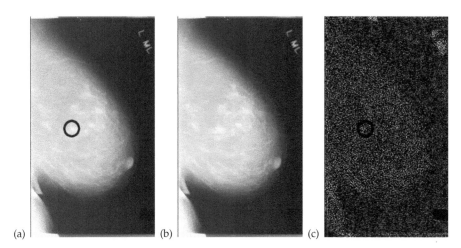

**FIGURE 6.12**
Changes introduced by jpeg compression at 50% quality level. (a) Original image, (b) post-compression image, (c) result of subtracting post-compression image from original image. (Original from the Digital Database for Screening Mammography [1,2]. With permission).

An example of the introduction of artifactual information into an image by jpeg compression is shown in Figure 6.12. It is for demonstration only: one wouldn't dream of performing lossy compression on a mammogram, as the abnormalities are too subtle and could easily be destroyed. A mammogram is shown Figure 6.12a with the abnormality (a calcification) circled. The image that has undergone jpeg compression, Figure 6.12b, looks visually similar. Figure 6.12c shows the result of subtracting the two. Artifacts are seen around the "LML" in the upper right of the image and there are some particularly bright spots inside the circle. Furthermore, the difference between the two images is not zero. The image saved using lossy jpeg compression has different pixel values from the original.

Jpeg effects are studied, using a specially designed test image, in a case study in Chapter 11.

### The JPEG2000 Format

JPEG2000 is a successor to JPEG and differs from it significantly because JPEG2000 supports both lossy and lossless compression. The DICOM standard includes compression using lossy JPEG, lossy and lossless JPEG2000 and lossless run length encoding; always ensure that a lossless method is used.

## Activity: JPEG Compression in ImageJ

The degree of jpeg compression is expressed in terms of the quality of the result. One hundred percent quality represents no compression, and decreasing percentage qualities are associated with higher compression ratios.

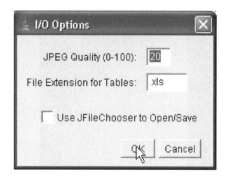

**FIGURE 6.13**

The ImageJ Input/Output options dialog, used to set the required quality level for saved JPEG images.

Start ImageJ and open the image file retinalimage.tif.*

Select Edit | Options | Input/Output.

A window similar to Figure 6.13 will appear. Change the value in the box labeled "JPEG Quality (0–100)" to 20 and click OK.

Select File | Save As | Jpeg ... and save the image as retinalimage_20.jpg. Close the image by clicking on the cross at the top right.

Use File | Open to open retinalimage.tif and then retinalimage_20.jpg. Move the images on the desktop until you can compare them visually side by side.

Click on each image in turn and use Image | Zoom | View 100%. This means that any additional artifacts arising from scaling the images for display will not be seen. The degradation in image quality, especially of linear features in the image, is very visible for this high degree of compression.

Close the jpeg image, but keep the tif image open.

Select Edit | Options | Input/Output. Change the value in the box labeled "JPEG Quality (0-100)" to 90 and click OK.

Select File | Save As | Jpeg ..., and save the image as retinalimage_90.jpg. Close the image by clicking on the cross at the top right.

---

* Image data from the DRIVE retinal image database, www.isi.uu.nl/Research/Databases/ DRIVE/. With permission.

Repeat the visual comparison using the _90.jpg image instead of _20.jpg. By eye, they look very similar, but the compression/decompression cycle has changed the value of almost every pixel in the image (test this by subtracting this pair of images using the image calculator in ImageJ). The effect of compression on pixel values is why jpeg compressed images must not be used in image analysis—they simply do not have the same pixel values as the original.

Look in Windows® Explorer and compare the sizes of the stored images. The original tif file has 323 KB compared with only 47 KB in the 90% compressed file. This compression ratio of over 6 is a clue that lossy compression was used.

Be aware too that:

- Image data formats are easily changed. Just because an image is presented in a format that avoids lossy compression, that doesn't mean that it has never been compressed.
- Every time an image is saved in a compressed jpeg format, the data are changed.

## 6.4 Good Practice Considerations

When working with medical images, it is essential to ensure that original images are available, together with all their header information.

If image compression is to be used, then lossless compression is the only suitable method for medical image processing.

When changes are made to image data, information on the changes made must be saved with the new images. Knowledge of image provenance is vital.

## 6.5 Chapter Summary

In this chapter the pros and cons of different image data formats were described, with particular reference to the special needs associated with medical image data. The concepts of image compression were introduced, and the importance of using lossless compression techniques in medical image analysis was emphasized. ImageJ activities allowed the reader to investigate related points in an interactive manner.

## 6.6   Feedback on Self-Assessment Questions

### Self-assessment question 6.01

A 16-bit image requires 2 bytes per pixel. Thus the file size required for the image data is 4000 × 5000 × 2 bytes, that is, 40,000,000 bytes or approximately 40 MB. This means that just one image would require 40 floppy disks, and a CD of 650 MB capacity would hold only 16 images.

### Self-assessment question 6.02

The images in Figures 6.9b and e are simple rotations of the original image; they do not incorporate the left-right reversal that occurs when the starting pixel is swapped from top to bottom. Images (c) and (d) do incorporate this; (d) is what would be loaded by the system, and (c) has had a further 180° rotation applied. A good way to think through this is to run a finger along the first row of the original image (we are told that the first row runs from top left to top right). The row goes first through the small circle and then through the triangle. The correct new image has its first row running from bottom left to bottom right. That first row will also go first through the small circle and then through the triangle. This happens for image (d).

### Self-assessment question 6.03

Applications that require lossless compression include primary diagnosis, archival images to be used for diagnosis, legal applications, image processing and analysis, teleradiology for diagnosis and product liability issues. Lossy compression would be appropriate for teaching files, as illustrations in reports, for general archiving (when diagnosis is known), for movie presentations and for presentation of research findings in articles and presentations. You may also have suggested that for some specific clinical applications, lossy compression may be considered appropriate; that regions of interest may be stored using lossless methods with the rest of the image using lossy methods; and that images in an electronic patient record might also be archived using lossy methods, if they are not to be used for diagnosis in the future. In any situation it is worth thinking through the implications of the choice before deciding to use lossy compression.

### Self-assessment question 6.04

The encoded segment is:

255|2   254|1   255|1   200|2   150|5   100|1   101|1   150|2   180|1   179|1   180|5   175|1   170|1

The original run required 24 bytes. The RLE run requires 26 bytes. The compression ratio is 0.92, and the value is less than 1 because the "compressed" file is larger than the uncompressed one. In this example there were several runs of only 1 pixel, which led to an increase in the size.

## References

1. Heath, M., et al. Current status of the Digital Database for Screening Mammography, in *Digital Mammography*, Karssemeijer, N., et al., Eds., Kluwer Academic Publishers, 1998, 457.
2. http://marathon.csee.usf.edu/Mammography/Database.html.

# 7

## Image Restoration

### 7.1 Introduction

The aim of image restoration is to correct imperfections introduced by the imaging system. Three kinds of imperfections are considered in this chapter: blurring introduced by the system response, geometrical distortion and gray-level inhomogeneity. Blurring introduced by the system may be partially removed by the process of deconvolution. The concept of the modulation transfer function, a plot used to indicate the degree of blurring of an imaging system in terms of the spatial resolution, is introduced. Correction of geometrical distortion and gray-level inhomogeneity is important when performing image analysis and synthesis to avoid errors affecting diagnosis, monitoring or therapy.

#### 7.1.1 Learning Objectives

When you have completed this chapter, you should be able to

- Define the point spread function (PSF) and modulation transfer function (MTF)
- Outline, in a nonmathematical manner, the procedure of deconvolution
- List the reasons that deconvolution will not result in the perfect image
- Use ImageJ to perform deconvolution
- Describe geometrical distortion and gray-level inhomogeneity and methods for their correction
- List image processing operations that would be adversely affected by the presence of geometrical distortion and gray-level inhomogeneity

#### 7.1.2 Example of an Image Used in This Chapter

*Self-assessment question 7.01*

Figure 7.1 is an image acquired using terahertz pulsed imaging, which is a method that is not yet widely used. The image shows a spatial resolution test

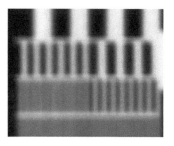

**FIGURE 7.1**
Image for self-assessment question 7.01. The image shows a spatial resolution test object acquired using terahertz pulsed imaging (Teravision project).

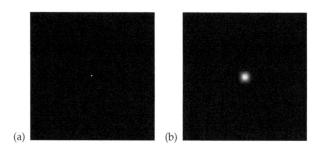

(a)       (b)

**FIGURE 7.2**
Illustration of the blurring effect of an imaging system. (a) Point object, (b) image of the point object.

object, consisting of several sets of high-contrast bars. Test objects are commonly used in imaging, and are made from different materials depending on the imaging modality under study. Notice how the closely spaced bars in the lower part of the image are not resolved from one another. What do you notice about the brightness of these bars compared with those at the top?

## 7.2 Blurring Arising from the Imaging System Itself

### 7.2.1 Point Spread Function (PSF)

Any imaging system has a blurring effect. If a point object is imaged, the resulting image is not a perfect point, but a spread-out version of that point, as illustrated in Figure 7.2. Mathematically, this blurring effect is described using the convolution operation (Chapter 4). Every imaging system has a characteristic blurring function, and the image of any object is the result of convolving the perfect image with this blurring function. The characteristic blurring function is called the *point spread function* (PSF). The PSF for an imaging system can be obtained theoretically from knowledge of the imaging process, or by practical measurement. In the latter case this means acquiring an image of an object with known properties, such as a point source or sharp edge.

Knowledge of the PSF of an imaging system is useful because of the potential for removing blurring by way of an inverse convolution operation, or deconvolution, between the measured image and the point spread function.

### 7.2.2   Deconvolution

In Chapter 4, the convolution theorem was introduced to explain how convolution filtering in the spatial domain can be replaced by multiplication in the frequency domain. The convolution theorem is also the reason that an efficient way to perform deconvolution is to work in the frequency domain, this time performing division between a pair of Fourier transforms rather than multiplying them.

The deconvolution process of an image with the point spread function of the imaging system does not result in the perfect image of an object because:

- Random noise is always added to an image by imaging devices. Thus image degradation is not purely as a result of convolution with the PSF.
- It is difficult to obtain a perfect estimate or measurement of the PSF, which adds further uncertainty to the calculation.
- Errors are amplified because the calculation involves division by small numbers.
- A theoretical PSF may contain zeroes, which is a problem as it will cause calculations to fail because of division by zero.

The ability of deconvolution to reduce blurring is shown in Figure 7.3. The result of straight deconvolution is less satisfactory when noise is present (Figures 7.3d – e), and it is usual to introduce some further frequency domain filtering to improve the result (Figure 7.3f).

Deconvolution can improve the appearance of images and it facilitates quantitation, so even though it does not completely remove the effects of the imaging system, the technique is widely used. Examples include:

- Scatter removal in SPECT
- Removal of out-of-focus information from microscope images
- Removal of the effect of recirculation of the radioisotope bolus when analyzing time activity curves
- Removal of system and atmospheric absorption effects from a measured terahertz pulse in terahertz pulsed imaging

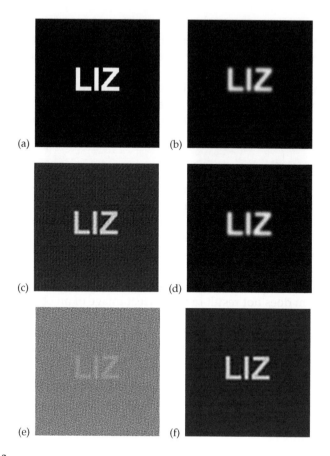

**FIGURE 7.3**
The reduction of blurring using deconvolution. (a) Original, (b) blurred image of the original, (c) result of deconvolution on the blurred image, (d) noisy and blurred image of the original, (e) result of deconvolution on the noisy and blurred image, (f) result of deconvolution on the noisy and blurred image, with frequency domain filtering to remove high frequencies.

### Activity: Effect of Small Differences in the PSF on the Deconvolution Result

In this activity, ImageJ is used to perform deconvolution using an image and an image of the system PSF. To simulate the variability that may arise in measuring the system PSF, a small alteration is made to the PSF. The effect on the result of deconvolution is viewed.

Load the image chapter7_blurred.tif from the CD.

Load the PSF image psf.tif from the CD.

Select Process | FFT | FD Math. Set up the dialog box for deconvolution as shown in Figure 7.4. Click OK.

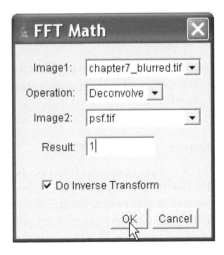

**FIGURE 7.4**
The ImageJ frequency domain math dialog.

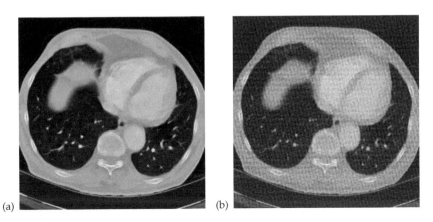

(a)      (b)

**FIGURE 7.5**
Results from the ImageJ activity to show the effect of small differences in the PSF on the deconvolution result. (a) Result using the original PSF, (b) result using PSF with added noise.

Save the result (Figure 7.5a) as 1.tif, and close image 1.tif.

Select psf.tif by clicking on it. Select Process | Noise | Add Specified Noise and set the standard deviation in the box to 0.5; click OK.

Perform deconvolution again, using the slightly revised psf file, and naming the result 2. Even this small change to the PSF has led to a marked deterioration on the result of deconvolution (Figure 7.5b). Reopen 1.tif for comparison if wished. Save the result as 2.tif; close image 2.tif and close psf.tif without saving.

If wished, use the ImageJ Image Calculator to subtract the new PSF from the original. Run the cursor over the result to display the pixel values.

Try changing the PSF using other processing functions, such as blur and sharpen. Avoid possible confusion by closing files each time, and reopening a fresh copy of psf.tif for editing.

### 7.2.3 Modulation Transfer Function (MTF)

The Fourier transform of the point spread function is called the *modulation transfer function* (MTF). The MTF is used in the deconvolution process, but it is also useful to be aware of this function for another reason. The MTF gives a graphical indication about the performance of an imaging system in terms of the relative amount of information that is retained about different spatial frequencies in an image. You may see MTFs presented in promotional material for imaging equipment, and it is necessary to take information only from the part of the curve that is at the relevant spatial frequency for the diagnostic application of interest.

Figure 7.6a is an image of a resolution test object acquired using an early terahertz pulsed imaging system. The bar pattern at the top of the image is clearly resolved, but it is hard to see the closely spaced bars in the lowermost row in the image. In other words, the high spatial frequencies are not resolved in this image. Profiles drawn through the four rows of bars on the test object are plotted in Figure 7.6b to emphasize this point. Note also how the contrast between the bars and the spaces between them decreases as the spatial frequency increases. The MTF for the system that produced the image is shown in Figure 7.6c. Here the same information that is present in the image is presented in a quantitative way. The theoretical limiting value of an MTF for resolution is 9%, but in practice it depends on the viewing conditions, and sometimes an observer may resolve a pattern where the MTF is as low as 2%. In this case, the MTF is under 9% for spatial frequencies corresponding with the two most closely spaced groups of bars on the test object.

*Self-assessment question 7.02*

Which of the following are reasons that deconvolution does not result in a perfect image?

A. Random noise contributes to image degradation in addition to the blurring from the system response represented by the PSF.

B. The convolution theorem is wrong.

C. Small errors in the PSF used for deconvolution can make a big difference to the image obtained.

D. Many software packages implement convolution filtering incorrectly.

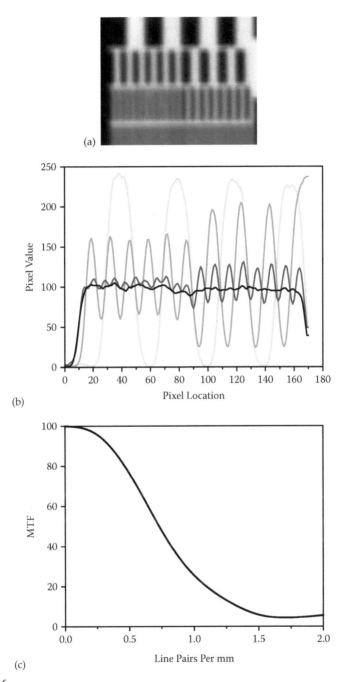

**FIGURE 7.6**
Measuring the reproduction of spatial frequencies in an image using a test object. (a) Image of a spatial resolution test object acquired using terahertz pulsed imaging, (b) intensity profiles for the four rows of bars. The profile for the most closely spaced bars is shown in black and for the most widely spaced bars in light gray. (c) The modulation transfer function (MTF) for the terahertz pulsed imaging system calculated using the information in the acquired image.

*Self-assessment question 7.03*

Figure 7.7a shows a blurred noisy image. Figure 7.7b shows the frequency domain spectrum after deconvolution of the image with a PSF. Three frequency domain filters are shown in Figures 7.7c–e, in which the black areas will set the value of the spectrum to zero and the white will leave the spectrum unaffected. Which filter would, after inverse transformation, result in the image in Figure 7.7f?

*Self-assessment question 7.04*

Sketch the MTF shown in Figure 7.6c. Add (i) a sketched curve representing an imaging system that has better spatial resolution than the original for all spatial frequencies above 0.2 line pairs per mm, and (ii) a sketched curve representing an imaging system that has slightly better spatial resolution than the original for spatial frequencies above 0.75 line pairs per mm, and slightly worse spatial resolution than the original for spatial frequencies below 0.75 line pairs per mm.

## 7.3 Geometrical Distortion

The two main types of geometrical distortion, barrel and pincushion, are illustrated in Figure 7.8.

### 7.3.1 Methods for Correction of Geometrical Distortion

Geometrical distortion is possible in all imaging modalities, which is why clinical imaging equipment undergoes regular quality assurance testing, so that the users can be sure that the system is performing within its prescribed limits. If the distortion was found to have increased, the system would be repaired before errors could occur. Image restoration discussed in this chapter involves correcting the effect on images of distortion that is present in a system, even when it is within the operating limits, so that it does not lead to errors affecting diagnosis, monitoring or therapy in:

- Measurements of area and volume
- Comparisons made with images acquired using other modalities
- Alignment of images acquired using other modalities or acquisition techniques

The correction procedure requires images of a phantom designed with components of known sizes and locations in the field of view. For 2-D imaging systems, this may be as simple as a grid, though a more complex arrangement is used in 3-D. A correction can be calculated that will result in

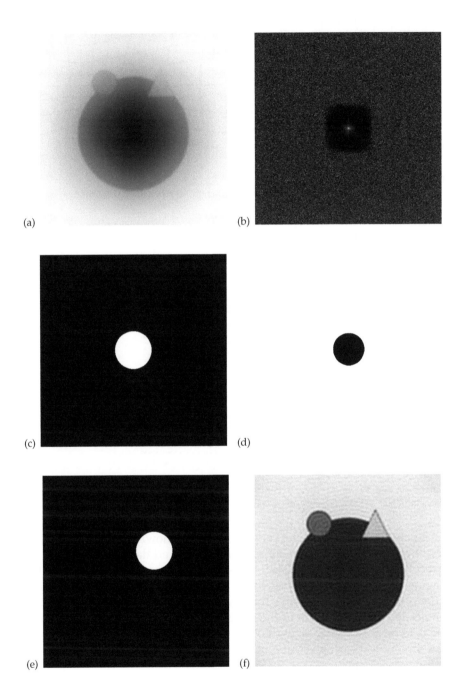

**FIGURE 7.7**
Images for self-assessment question 7.03. (a) Blurred and noisy image, (b) frequency domain spectrum associated with the image after deconvolution, (c)–(e) frequency domain filters, (f) image after frequency domain filtering.

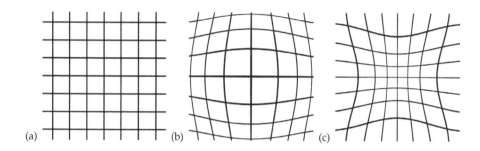

**FIGURE 7.8**
Geometrical distortion. (a) Original, (b) barrel distortion, (c) pincushion distortion.

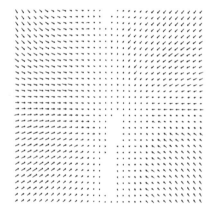

**FIGURE 7.9**
A correction map to remove the barrel distortion from the image of Figure 7.8b. The array of arrows indicates the magnitude and direction of the corrections.

coordinates in an image being returned to their known, true locations. That correction is then applied to subsequently acquired data. For example, the correction shown in Figure 7.9 is the one that would be applied if the result of imaging the square grid was the barrel grid image of Figure 7.8. Each arrow in the correction map indicates the magnitude and direction of the correction needed at that location in the image. Another way to show this kind of data is to have two images, with one showing the required displacement in the x direction at each pixel, and the other showing the same information for the y displacement.

*Self-assessment question 7.05*

Figure 7.10a shows a test pattern that will be imaged. Figure 7.10b is the result of imaging the test pattern. Does this result illustrate barrel or pincushion distortion?

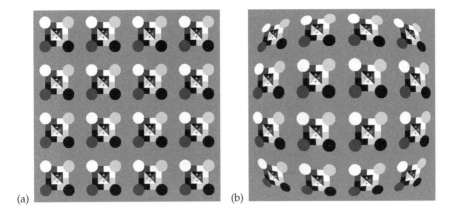

(a)                                    (b)

**FIGURE 7.10**
Images for self-assessment question 7.05. (a) Test pattern, (b) image of test pattern.

## 7.4   Gray-Level Inhomogeneity

Gray-level inhomogeneity is illustrated for different imaging modalities in Figure 7.11. When inhomogeneity is present, areas of the image representing parts of the object with the same properties do not have identical image brightness as would be expected.

### 7.4.1   Methods for Correction of Gray-Level Inhomogeneity

As is the case for geometrical distortion, gray-level inhomogeneity is monitored under the routine quality assurance testing undergone by clinical imaging equipment. Correction of inhomogeneity is not routinely performed, but is necessary when image analysis is to be performed that relies on consistent values across the field of view. Examples include image segmentation or classification techniques that rely on gray level, and the calculation of enhancement from the presence of contrast medium. The correction is sometimes called *bias correction*. There are two broad approaches.

#### 7.4.1.1   Correction Using an Image of a Uniform Object

The first approach is similar to the method described previously for correction of geometrical distortions: measurements are made from a uniform object and these are used to correct acquired data. The uniform object, which supplies a *flood* or *flat field*, is usually a liquid-filled cylinder. The disadvantage of this approach is that the uniform field should be acquired immediately before or after the patient scan. This means that the need for a correction must be recognized in advance of the first patient study. The method cannot be applied retrospectively to previously acquired scans that

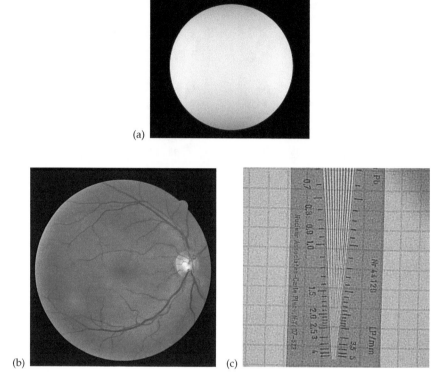

**FIGURE 7.11**
Gray-level inhomogeneity seen in different imaging modalities. (a) Magnetic resonance imaging, (b) retinal imaging, (c) x-ray image intensifier. (Courtesy of Sarah Bacon, www.isi.uu.nl/Research/Databases/DRIVE/[1] and GE Healthcare. With permission.)

lack the associated uniformity acquisitions. Furthermore, this approach does not correct for subject-dependent effects.

### 7.4.1.2  Correction Derived from Acquired Subject Data

The second approach uses the acquired subject data themselves. This approach has the advantage of being possible retrospectively, and of including subject-dependent effects. A range of algorithms for applying this technique has been investigated. The simplest algorithms involve low-pass filtering of the image, and result in an image which approximates the image acquired from a uniform phantom. All the image detail is blurred out, but the large-scale intensity variation across the image is retained. More complicated methods rely on identifying areas of the image representing the same tissue, which should have the same signal intensity, and deriving a correction map from this information.

## Activity: Simple Inhomogeneity Correction Using ImageJ

In this activity, artificial inhomogeneity is first added to an image. This image is then used to demonstrate the principles of inhomogeneity correction in an exaggerated case of inhomogeneity.

Load the image slice98.tif* from the CD.

Select File | New using the values shown in Figure 7.12.

Select Image | Rotate | Rotate 90 Degrees Right.

Select Process | Image Calculator, and use it to add the two images. Select a 32-bit result to avoid saturation.

Click on the result and convert it back to 8 bits using Image | Type | 8-bit. Save the image as slice98plusramp8bit.tif.

This image (Figure 7.13a) clearly exhibits intensity inhomogeneity. To demonstrate how thresholding on gray value would lead to useless results, select Image | Adjust | Threshold. Adjust the range using the sliders and note how the selected red areas of the image do not match tissue types.

Click on Reset, then close the threshold window.

Click on the image you made, then make a duplicate by selecting Image | Duplicate. If you accept the name offered, this will be called slice98plusramp8bit-1.tif

Select Process | Filters | Gaussian Blur and choose a radius of 53. This makes a very blurred version of the inhomogeneous image. All the detail has gone, but the top-to-bottom inhomogeneity is still visible (Figure 7.13b).

Select Process | Image Calculator and subtract the blurred image from slice98plusramp8bit.tif. Create a new window and a 32-bit result to allow for results not in the range 0–255.

Click on the result and convert it back to 8 bits using Image | Type | 8-bit. Save the image as slice98plusramp8bitcorrected.tif.

Select Image | Adjust | Threshold. Adjust the range using the sliders and note how the selected red areas of the image match tissue types much better than they did before. Click on Reset, then close the threshold window.

---

* Image data from the BrainWeb Simulated Brain Database [2,3]. With permission.

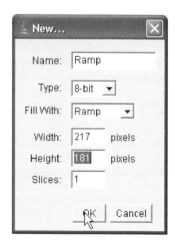

**FIGURE 7.12**
The ImageJ New Image dialog.

Close all the open images except the original slice98.tif.

Select Image | Adjust | Threshold. Adjust the lower slider to 255, and move the upper slider from left to right to about 200. Although far less obvious than our artificially added ramp, there is clearly inhomogeneity present in this image (Figure 7.13c)

Repeat the correction on this image and compare the symmetry of thresholding the result.

The value of 53 was chosen for the radius of the Gaussian blur because it was larger than the size of most of the features in the image. Experiment with other values.

*Self-assessment question 7.06*

Which of the following are true statements regarding gray-level, or intensity, inhomogeneity in an image?

A. You can recognize intensity inhomogeneity by observing that the same tissue is represented by very different gray levels in different parts of the field of view.

B. Intensity inhomogeneity can be corrected by acquiring an image of an object known to be uniform and applying a correction derived from this to the measured data.

C. Intensity inhomogeneity can be corrected by using data from the acquired image itself; one option is to blur the image strongly to generate a correction field.

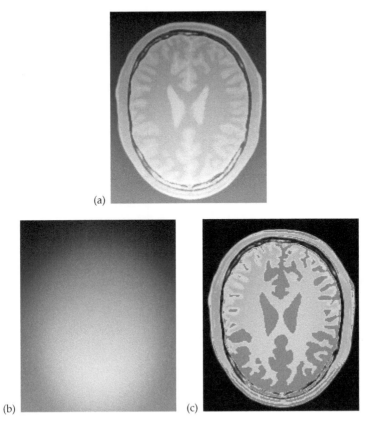

(a)

(b)

(c)

**FIGURE 7.13**
Images associated with the inhomogeneity correction activity. (a) Image with added inhomogeneity, (b) highly blurred version, (c) thresholding the original image shows that inhomogeneity is present. The dark gray areas in the image will appear in red in ImageJ. (Image data from the BrainWeb Simulated Brain Database [2,3]. With permission.)

    D. Gray-level inhomogeneity is only seen in MR images.

    E. Gray-level inhomogeneity does not affect the performance of clustering algorithms.

*Self-assessment question 7.07*

Which of the images in Figure 7.14 show gray-level inhomogeneity and which geometrical distortion?

*Self-assessment question 7.08*

There is an option in the Process Menu of ImageJ called Subtract Background. Read the documentation to determine the recommended size of the "Rolling Ball Radius." What are the implications about the type of image that

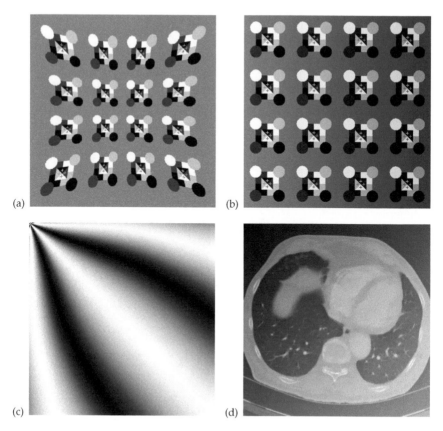

(a)                                              (b)

(c)                                              (d)

**FIGURE 7.14**
Images for self-assessment question 7.07. The images all have added geometrical distortion or inhomogeneity. For comparison, similar images without the added effects may be seen in (a) Figure 7.10a, (b) Figure 7.10a, (c) Figure 1.7a, and (d) Figure 7.5a.

is expected? Apply the filter to slice98plusramp8bit.tif (which is the highly inhomogeneous image you saved during the last activity). Use a Rolling Ball radius based on your reading of the documentation, ensuring that the white background option is not selected. Did the operation have the desired effect? What other information would you need to know about this option before including it in an image processing protocol?

## 7.5    Good Practice Considerations

Geometrical and gray-level corrections that have been performed on an image must be fully recorded and reported as they form part of the image analysis procedure.

Deconvolution must be performed using a measured or theoretically predicted point spread function. There is no justification for creating a pseudo-PSF without underpinning measurement or theory, to use in the deconvolution process.

Ensure that the parameters of any correction functions provided in software packages are fully understood before use.

## 7.6 Chapter Summary

This chapter covered image restoration, which is concerned with the correction solely of imperfections introduced by the imaging system. In the section on image blurring, the point spread function and the concept of deconvolution were introduced. ImageJ was used to demonstrate how sensitive the deconvolution result can be to small differences in the input functions. Methods for correcting geometrical distortion and gray-level inhomogeneity were outlined.

## 7.7 Feedback on Self-Assessment Questions

### Self-assessment question 7.01

The difference in intensity between the bright and dark parts of the bar pattern, or image contrast, is very much reduced for the closely spaced bars compared with the widely spaced ones. This difference in the modulation of the signal is discussed in Section 7.2.3.

### Self-assessment question 7.02

Reasons A and C are reasons that deconvolution does not result in a perfect image. The convolution theorem is not wrong (B). Although it is true that many software packages wrongly implement convolution filtering by using the correlation operation (D), this does not affect the frequency domain operation of deconvolution.

### Self-assessment question 7.03

The correct filter is the one shown in Figure 7.7c. This filter will remove the high-frequency noise in the spectrum, but retain the image detail at the center. The filter in Figure 7.7d would retain only the noise, and the filter in Figure 7.7e does not include the important low-frequency information at the center of the spectrum.

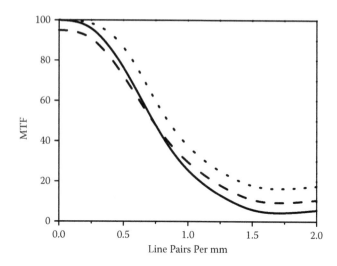

**FIGURE 7.15**
Feedback on self-assessment question 7.04. The dotted line is the sketch for condition (i), and the dashed line is for condition (ii).

## Self-assessment question 7.04

An example of the appearance that should be achieved is shown in Figure 7.15, where the dotted line is curve (i) with better spatial resolution than the original for all spatial frequencies above 0.2 line pairs per mm, and the dashed line is curve (ii). In practice you are more likely to perform this process in reverse, and be able to identify which curve on an MTF represents the higher resolution for a given spatial frequency.

## Self-assessment question 7.05

The distortion shown in Figure 7.10b is barrel distortion.

## Self-assessment question 7.06

Statements A, B and C are all true. Although gray-level inhomogeneity is a problem in MR images, MR is not the only imaging modality to suffer the effect. Clustering algorithms would be strongly affected by the presence of gray-level inhomogeneity, as they are based on the premise that similar materials are imaged with similar intensity.

## Self-assessment question 7.07

Figures 7.14a and c show geometrical distortion (pincushion and barrel, respectively). Figures 7.14b and d show gray-level inhomogeneity.

## Self-assessment question 7.08

The documentation states that the Rolling Ball Radius should be at least as large as the radius of the largest object in the image that is not part of the background. This implies that the kind of image expected is one similar to a microscope image or gel, where several objects are distributed across a background. The documentation talks about a smooth continuous background, and while this may describe intensity nonuniformity in some radiological imaging modalities, it is not clear that the Subtract Background operation will be applicable. Clearly, before using any option in a software package, it is necessary to find out exactly what is being done. It is expected that, as is the case here, there is reference in the software manual to an original article on the method. Having determined the necessary information, it will then be possible not only to justify use of the method, but also to report it properly so that others may repeat the work.

## References

1. Staal, J.J., et al., Ridge based vessel segmentation in color images of the retina, *IEEE Trans.Med. Imag.*, 23, 501, 2004.
2. The BrainWeb Simulated Brain Database. www.bic.mni.mcgill.ca/brainweb/.
3. Collins, D.L., et al., Design and construction of a realistic digital brain phantom, *IEEE Trans. Med. Imag.*, 17, 463, 1998.

# 8

## Image Registration

## 8.1 Introduction

Image synthesis is the general term for bringing together information from more than one image. Synthesis can be separated into two parts: image *registration*, which covers the processes required to bring images into spatial alignment, and *visualization*, which allows information from the aligned data sets to be viewed together. Reasons for performing image synthesis include:

- Assessment of disease progression or growth using temporal series of images
- Combination of structural and functional information from different imaging modalities
- Comparison of corresponding regions in different individuals by matching both to a standard coordinate system
- Generation and analysis of atlases or templates representing the typical appearance in health or disease
- Arithmetical or statistical operations, such as subtraction in contrast enhanced data or statistical parametric mapping, which require registered images
- Providing a roadmap for invasive procedures and image guided surgery

### 8.1.1 Learning Objectives

When you have completed this chapter, you should be able to

- List several applications for image synthesis
- Define the four steps of image registration
- Define and distinguish between the four types of image transformation
- Suggest alternative methods of visualization of synthesized images
- Use ImageJ to perform image registration

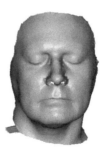

**FIGURE 8.1**
An image from a laser surface scanning system for self-assessment question 8.01

### 8.1.2   Example of an Image Used in This Chapter

*Self-assessment question 8.01*

Figure 8.1 shows an image generated from a laser surface scanning system. Laser scanning is a good way of measuring the shape of the skin surface without an ionizing radiation hazard. List two potential applications areas for this type of scanning.

## 8.2   Image Registration

### 8.2.1   Overview of the Four Steps of Image Registration

There are four general steps involved in image registration:

1. Feature extraction
2. Pairing
3. Calculation of transformation
4. Application of transformation

These are outlined briefly here, and then each step is considered in more detail. The general process is to identify *features* that are present in the two images. The features may be specially placed markers, or be image characteristics of a more subtle nature. The *pairing* process then determines which feature in one image should be aligned with which feature in the other. This is easiest to envisage when the markers have been placed deliberately, and it is simply a case of identifying the same markers in each image. In the *calculation* step, the mathematical operation that would be necessary to align the sets of paired features is found. Finally, in the application of the *transformation* step, the result of the calculation is applied to all the pixels or voxels in one of the image sets to align it with the other. The image data set which is transformed is called the *match,* and the one to which it is aligned is the *base* set. We will see that in some cases where numerous features are used, steps 2 and 3 are very closely related.

## 8.2.2 Step 1. Feature Extraction

One needs to identify features that appear in both images. The simplest way to approach this is by deliberately placing markers that show up well in the particular imaging modality. This is why early work used stereotactic frames that were imaged along with the brain of the subject. These frames provided very clearly identifiable features to aid navigation; the markers are known as *fiducials,* which in this context means a fixed standard.

However, the need to place markers is very restrictive because it means that registration can only be performed when there has been a plan to do so from the outset. Marker placement may also be invasive, in some cases requiring placement under anesthetic and with the potential to cause nerve injury. There is a trend for natural features that form anatomical landmarks to be used instead of specially fixed markers. Such landmarks may be:

- Anatomical landmarks selected manually in image data
- Anatomical landmarks defined by image analysis
  - The boundary of a surface (such as the brain or skull)
  - Computer-recognizable points such as the tips of the fingers

A further refinement is to use numerous features. These could be the values of the pixels themselves or quantities derived from the image data such as texture measures. The large number of features means that the methods used for the pairing and transformation steps of registration are different from those used for features derived from fiducials or anatomical landmarks. The two kinds of features are discussed separately in the following sections.

## 8.2.3 Step 2. Pairing

Once features have been identified in both the images, they need to be analyzed to determine which feature in one image matches which feature in the other. Pairing is necessary so that, for example, the tip of the nose is aligned with the tip of the nose, rather than the tip of the nose with the point of the chin. This process of pairing is also called identifying *correspondences.*

### 8.2.3.1 *Identifying Correspondences where Fiducials or Anatomical Landmarks Are Used as Features*

The principle is illustrated for three landmarks in Figure 8.2. In each of the two images (a) and (b), the white dot represents the center of the right eye, the

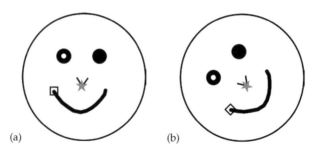

(a)                                                    (b)

**FIGURE 8.2**
(a)–(b) Corresponding fiducials, or anatomical landmarks, in two misregistered images. The white dot represents the center of the right eye, the gray star is at the tip of the nose and the open black square is at the end of the mouth.

gray star is at the tip of the nose and the open black square is at the end of the mouth.

### 8.2.3.2  Identifying Correspondences where the Pixel or Related Values Are Used as Features

The term *similarity metric* is the general way of describing a measure that is used in the process of pairing points with similar properties. An example of a similarity metric is the correlation coefficient. Imagine taking two images that are identical apart from the fact that they are misaligned. One image is fixed in place, and the other one is moved into all possible positions and orientations that it might take. At each location, each pixel value in the fixed image is multiplied by the value of the pixel that is overlying it, and the results for all the pixels are summed. The sum will be greatest when the two images are in perfect alignment. This process is called *correlation*, and the correlation coefficient (R) is unity for two perfectly aligned images. Correlation is illustrated in the diagram in Figure 8.3. Images (b)–(e) have each been correlated with image (a). The corresponding *joint histogram* is shown for each pair in (f)–(i). The joint histogram is a plot of the pixel value in one image against the pixel value in the other. For two identical images in perfect alignment, the plot is a straight line. The square of the correlation coefficient ($R^2$) is given in the caption. As the misregistration increases from (b)–(e), the value of $R^2$ goes down. Correlation can therefore be used to identify correspondences, by finding the point pairs that result in the maximum value of the correlation coefficient. This works when the images are of the same modality, and thus have the same relationship between voxel values. Note that the general term to describe techniques where the same procedure is repeated, with the starting conditions updated according to the outcome of the previous round of calculations, is *iterative*.

A related similarity metric is *mutual information*, which is maximized to achieve image registration. The maximization of mutual information method is suitable even if the images are from different modalities and so may have widely differing pixel values representing the same object in the

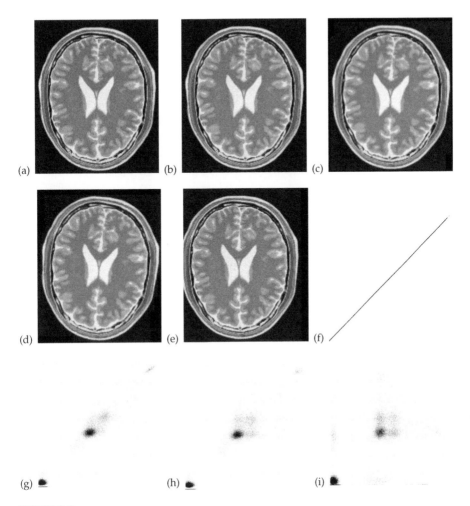

**FIGURE 8.3**

(a) – (e) Images, increasingly misregistered with image (a). (f) Joint histogram for images (b) and (a), $R^2 = 1.0$. (g) Joint histogram for images (c) and (a), $R^2 = 0.78$. (h) Joint histogram for images (d) and (a), $R^2 = 0.42$. (i) Joint histogram for images (e) and (a), $R^2 = 0.38$. (Image data from the BrainWeb Simulated Brain Database [1,2]. With permission.)

two images. The aim is to find the alignment, or pairing of points, which maximizes a mutual information criterion. Unlike the case for images from the same modality, for two images of the same object in different modalities, even when the images are aligned, there will be a spread of pixel values in the joint histogram instead of a straight line. However, it is the case for both single and dual modality cases that if the images are out of alignment, then the pattern of the joint histogram is less ordered. Maximization of mutual information is a method of registration that is based on finding the alignment, or pairing of points, that makes the histogram as *ordered* as possible. Another technique uses minimization of entropy for the same ends.

The features used for pairing need not be pixel values. Surface matching pairs a large number of points known to lie on each surface. Other methods include information such as the surface curvature, or distance from an edge, associated with each point.

*Self-assessment question 8.02*

The following are extracts from the abstracts of three different articles in which image registration is performed. One method pairs anatomical landmarks, one pairs points on a surface and one pairs pixel values. From the information provided in the extract alone, which pairing method is associated with each extract?

> "A surface matching technique has been developed to register multiple imaging scans of the brain in three dimensions, with accuracy on the order of the image pixel sizes."[3]

> "This method does not require fiducial markers and the user is not required to identify common structures on the two image sets. To align the images, the algorithm seeks to minimize the standard deviation of the PET pixel values that correspond to each MRI pixel value."[4]

> "The technique is based on user identification of point-like landmarks visible in both modalities."[5]

Often a rough preliminary registration procedure is performed first to bring the images into approximate alignment, before an attempt is made to identify the correspondences by the iterative methods previously described. This preliminary step is necessary because when there is a large misalignment between the images some methods fail (Section 8.2.6). Once preliminary alignment has been performed, it is usual to start the iterative calculations using just a few points, work out the arrangement that gives the most ordered joint histogram and then use that information as the starting value for the next set of iterations, using more points. The similarity metric may be calculated only for a subset of points in the center of the field of view, to exclude background regions.

## 8.2.4 Step 3. Calculation of Transformation

The number of pairs of features involved in registration can vary enormously depending on the type of landmark. A minimum of three pairs of fiducials is needed, in the case of manually identified landmarks usually 10 or more are used. If a surface has been defined computationally then of the order of 1000 is normal, and if voxel values are used then maybe 10 million point pairs are involved.

Calculations are performed to determine what mathematical operation (called a *transformation*) will give the best alignment of all the pairs of features when applied to the match set to align it with the base set. Four different types of transformation can be applied.

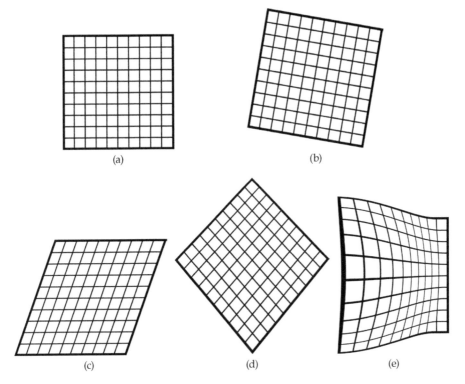

**FIGURE 8.4**
Illustration of the four types of geometrical transformation. (a) Untransformed grid, (b) rigid transformation, (c) affine transformation, (d) projective transformation, (e) curved transformation.

### 8.2.4.1 Types of Geometrical Transformation

There are four types of geometrical *transformation*, and these are shown in Figure 8.4:

- *Rigid*—In a rigid transformation all parts of the object are assumed to move as a whole, and there is no movement of one part of the object relative to any other. This may be appropriate to a solid object like the human skull, but not a deformable one such as the liver. A rigid transformation could be a translation (linear movement) as indicated in Figure 8.5b, where the initial position is shown dashed, or a rotation as seen in Figures 8.5c and d.

- *Affine*—In an affine transformation, straight lines remain straight and parallel lines remain parallel. Angles can change, too, as happens in a scaling (Figure 8.6b) or shearing (Figure 8.6c) deformation. Note that the rigid transformation is a subset of the affine transformation.

- *Projective*—Projective transformations (Figure 8.7) include further deformation. Straight lines remain straight but in projective transformations parallel lines need not remain parallel.

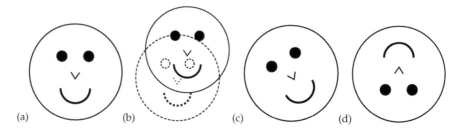

**FIGURE 8.5**
Examples of rigid transformations. (a) Original image, (b) translation, with initial size and position shown dashed, (c) rotation, (d) rotation through 180°.

**FIGURE 8.6**
Examples of affine transformations. (a) Original image, (b) scaling transformation, (c) shearing transformation.

**FIGURE 8.7**
Examples of projective transformations. (a) Original image, (b) and (c) projective transformations.

**FIGURE 8.8**
Examples of curved transformations. (a) Original image, (b) and (c) curved transformations.

- *Curved or elastic*—Curved or elastic transformations (Figure 8.8) are most appropriate for deformable tissue. In contrast with all the other transformations, straight lines need not remain straight. The curved or elastic transformation is the most general case of transformation. It should be used with care, and never applied to structures known to be rigid.

Note that these transformations are often further grouped into rigid versus nonrigid (i.e., all the transformation types except for the rigid one). The computational simplicity and the ease of evaluation means that rigid registration is very commonly used, even when a small amount of nonrigid deformation is expected.

The method used to calculate the transformation usually falls into one of two groups. The first applies where fiducials or anatomical landmarks are used as features, and the second where pixel values are used as features.

### 8.2.4.2  Calculation of the Transformation where Fiducials or Anatomical Landmarks Are Used as Features

For a small number of landmarks, the process of calculating the transformation is analogous with using the least-squares technique to fit the best straight line to a set of points on a graph. In that case, the aim is to fit a line with the smallest possible mean-squared distance between the measured points and the fitted line. In a similar way, the aim of the transformation calculation is to find the transformation that gives the smallest mean difference between the locations of pairs of points on the transformed match and the base set. The transformation that gives the closest match over the complete set of pairs will be found, so it should be noted that it is unlikely that every feature will be perfectly lined up with its partner. The calculated transformation will be applied to the complete data set in Step 4.

### 8.2.4.3  Calculation of the Transformation where Pixel or Related Values Are Used as Features

Where pixel values are used as features, the pairing process of Step 2 will include an estimate of the required transformation for the data set, because every possible alignment tested in the iterative process of pairing has a known transformation associated with it. The transformation will be applied to the complete data set in Step 4.

The overall iterative process is illustrated in the diagram in Figure 8.9. This diagram applies to cases where the pairing of points is done by trying many different arrangements to see which gives the best result in terms of the similarity metric. It does not apply to registration using paired landmarks where the pairing is well understood, such as with fiducials and anatomical landmarks.

*Self-assessment question 8.03*

Does Figure 8.10b or Figure 8.10c show a shearing deformation of the shape in Figure 8.10a?

*Self-assessment question 8.04*

The following is an extract from the abstract of an article [6] in which image registration is performed. Which of the four types of transformation is used in this work?

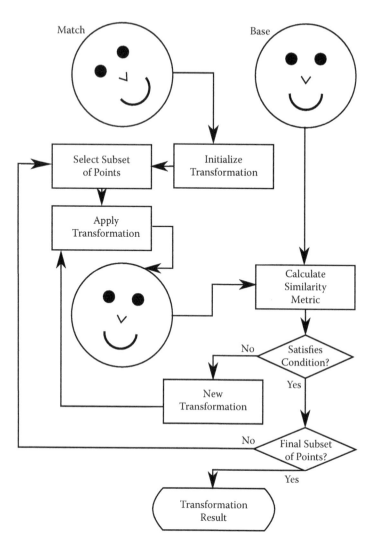

**FIGURE 8.9**
Flowchart showing the overall iterative registration process.

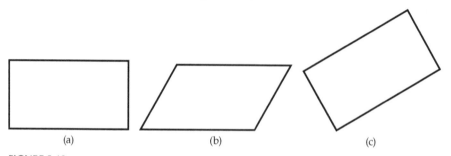

**FIGURE 8.10**
(a)–(c) Images for self-assessment question 8.03.

"T2-weighted axial magnetic resonance images and transaxial 99mTc-HMPAO and 201Tl images acquired with an annular gamma camera were merged using an objective registration (translation, rotation and rescaling) program."

*Self-assessment question 8.05*

The original image is Figure 8.11a. Match Figures 8.11b–e to the correct transformation from the following list: rigid, affine, projective, curved.

## 8.2.5 Step 4. Application of Transformation

The calculated transformation is applied to all the voxels in one of the data sets (the match) to align it with the other (the base). This is often called a *matrix transformation* or *matrix operation*. Matrix operations are mathematical operations and available in many image processing packages. Fortunately, a full understanding of matrix algebra is not required to use an image registration package successfully. However, one aspect that is important is the issue of interpolation (Chapter 1). Applying the transformation will often require pixel values in the match image to be interpolated. Interpolation can be thought of as filling a location where no pixel value was previously defined. For example, if the image is rotated and rescaled, new values must be calculated at the pixel locations of the base image, as illustrated in Figure 8.12.

## 8.2.6 Factors That Can Cause Errors in the Registration Process

Several factors must be kept in mind when performing image registration:

- All image registration routines are based on assumptions made about the data that are being aligned. For example, it is assumed that the two data sets are not severely distorted compared with one another. This assumption applies even for elastic transformations. Image preprocessing may be applied to correct for distortion effects before registration is attempted. Such distortions could arise from:
  - Grayscale inhomogeneity in the imaging process
  - Spatial distortion due to the imaging process
  - Patient movement
- It is often assumed that there is a corresponding point in the base volume for every point in the match volume; that is, that the match volume is totally included in the base volume. This condition may not be met if the fields of view for the two acquisitions differ considerably. In practice, when using software, one should take care to ensure that the base volume is chosen to be the one that covers the larger volume, and that the match volume is trimmed if necessary.

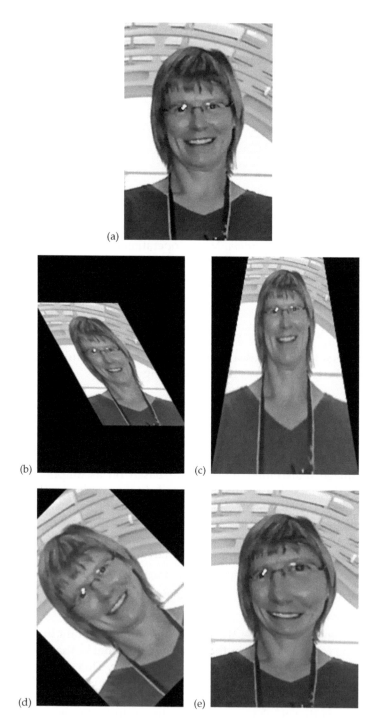

**FIGURE 8.11**
(a)–(e) Images for self-assessment question 8.05.

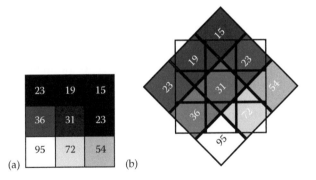

**FIGURE 8.12**
Interpolation and image transformations. (a) Image before transformation, (b) rotated and scaled image, showing pixel locations of the base image at which new values must be calculated.

**FIGURE 8.13**
Illustration of local and global minima.

- Some software packages are able to perform rigid transformations and not affine ones. If the voxel size is different in the two volumes, then they cannot be aligned with such software unless the scaling factor is first determined and applied.

- If fiducial markers smaller than the voxel dimensions are used, then there will be errors in marker localization. Thus, large fiducial markers will be used for low-resolution techniques, but this means there will be uncertainties in localization for the higher-resolution modality where use of a smaller fiducial would have been possible.

- Several techniques rely on the identification of a maximum or minimum value, for example, of the correlation coefficient or of mutual information. However, maxima and minima may be either *local* or *global*. This is illustrated in the diagram in Figure 8.13. Several points on the curve are mathematical minima (or turning points), but only one is the lowest valued minimum, which is known as the global minimum. The occurrence of local maxima and minima is a particular problem in medical imaging when rotational symmetry is present. For example, foreheads may appear to be well aligned in several different orientations each corresponding with local minima.

## Activity: Image Registration with Anatomical Landmarks Using ImageJ

Start up ImageJ and look in the Plugins menu to check that there is an option listed called TurboReg. If it is not present, refer to the installation instructions that precede Chapter 1.

Two images, chapter8_1.tif and chapter8_2.tif, are supplied on the CD. Load the images, taking care to load chapter8_2.tif before chapter8_1.tif. Now follow the steps below to align the two images.

Look at the images and decide on about six anatomical landmarks that you believe you can identify reproducibly on each image. Examples are the tip of the nose, clear shape features (such as turning points on outlines) or junctions between features. The minimum number of landmarks that is required depends on the type of transformation that is to be performed. The simplest is for a rigid body transformation, where a line between two points is required to define the angle of rotation and a single point (which does not lie on this line) to define the translation. For this activity, a good line to choose would run from the tip of the nose to a turning point or junction between features at the posterior brain. The single point can be any point away from this line that is clearly identifiable in both images.

Select Plugins | TurboReg | TurboReg.

A dialog box (Figure 8.14) will open; select the options shown. Do not press any of the buttons yet; instead, ensure that the two image windows are visible. Each should have blue, brown and green graphics overlaid. In chapter8_1.tif, click and drag the blue dot onto the tip of the nose, then drag the brown dot to the chosen point at the posterior brain. Drag the green cross to your chosen point away from the center line. Click to select chapter8_2.tif and repeat. (Figures 8.15a–c for chapter8_1.tif and (d)–(f) for chapter8_2.tif).

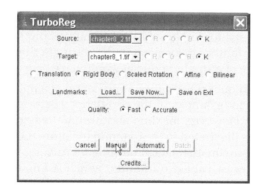

**FIGURE 8.14**
The ImageJ TurboReg dialog. (From http://bigwww.epfl.ch/thevenaz/turboreg/. [7] With permission.)

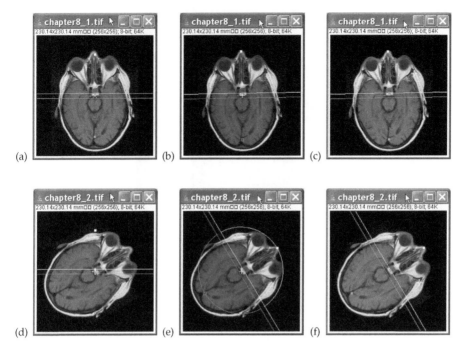

**FIGURE 8.15**

Positioning the landmarks in the TurboReg activity. (a) Image chapter8_1.tif before landmarks are moved, (b) chapter8_1.tif with blue square moved to tip of nose, (c) chapter8_1.tif with all three landmarks positioned, (d) chapter8_2.tif before landmarks are moved, (e) chapter8_2.tif with blue square moved to tip of nose, (f) chapter8_2.tif with all three landmarks positioned. (From http://bigwww.epfl.ch/thevenaz/turboreg/. [7] With permission.)

When the markers have been placed in both images, click on the Manual button. A new window labeled Output appears. This is a stack* of two images; the first image is chapter8_2.tif registered to chapter8_1.tif (Figure 8.16). This is why the loading order was important. The image termed the "source" image in TurboReg is the one that will be changed (the match image) and should be loaded first.

Select Image | Stacks | Convert stack to images. This converts the stack to two separate images. Select the registered image, select Image | Type | 8-bit, then save the registered image.

Use Process | Image Calculator (Subtract) to generate an image that shows the difference between the registered image and the base image chapter8_1.tif. Be sure to select the option for a 32-bit result. The result allows a visual assessment to be made of the registration result. This type of qualitative assessment is discussed in Chapter 10.

---

* You can read about stacks in the Image | Stacks section of the ImageJ documentation.

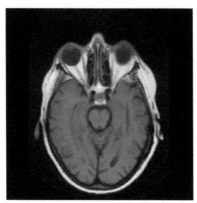

**FIGURE 8.16**
The result of the TurboReg activity, chapter8_2.tif registered to chapter8_1.tif.

*Self-assessment question 8.06*

If a pair of images were identical, or perfectly registered, which of the following gray values would you expect to see in the image that results from subtraction: −1024, −128, 0, +128, +1024?

### More Image Registration

If you enjoyed registering the images in the activity, investigate further the possibilities offered in the TurboReg plugin. For example, compare the options for scaled rotation or affine transformation with the rigid body transformation, paying attention to the number of landmarks used, or generate your own test misregistered images by using the Image | Rotate menu. There is another opportunity to perform image registration in Chapter 11.

## 8.3   Dimensionality and Number of Modalities

An important classification associated with registration is the *dimensionality*, which concerns the number of dimensions associated with the two data sets being aligned:

- 3-D to 3-D
  - Alignment of two sets of CT slices of the head.
  - Intraoperative alignment of a patient to preoperative MRI image slices.
- 2-D to 3-D
  - Projection to volume. Portal imaging in radiotherapy gives a low contrast 2-D projection image, which may be registered to a volumetric data set.
  - Single section from one 3-D acquisition to a volume from another 3-D acquisition.

- 2-D to 2-D.
  - A time series of images of a slice from a 3-D volume, where the slice is known to have a fixed plane. The time series images may need registration within the slice.
  - A set of slices exists that should be aligned to stack up to create a 3-D volume, but the necessary between-slice alignment has been lost, and the slices need to be aligned to create a credible stack. This situation is not common in CT/MR, but does occur for histological slices or microscope images.

Registration may also be classified in terms of the number of imaging modalities involved, that is, as *single modality* or *multimodality*. This is as straightforward as it sounds: if there is only one type of image involved, then it is a single modality example; if more than one, then it is multimodality.

*Self-assessment question 8.07*

Assign the following imaging tasks to either single or multimodality registration: assessment of disease progression, monitoring patient response to treatment, determination of the anatomical location of functional data.

---

## 8.4 Visualization of Registered Images

A further step is to display, or visualize, the two registered images. This is not strictly part of the process of registration, which is why it is considered separately and does not form the fifth step in the registration process. In some situations, the registration information may be derived and applied without any intention of displaying the two sets of data together. An example is in the calculation of contrast enhancement, where spatial alignment is key to the process, but the outcome of the contrast enhancement calculation is displayed rather than the two sets of aligned images.

In most situations the whole point of the registration is to make the best use of information from two sources by combining their displays. There are many ways of visualizing information from aligned data sets:

- Linked cursor display. Here the aligned images are shown separately. When the cursor is moved in one image, the cursor in the other image is simultaneously moved to the corresponding position.
- Color overlay display. One image, usually the one showing the anatomical features, is shown in gray scale with a color overlay of the functional image. The overlay may be opaque and completely replace the grayscale image, or it may be transparent, allowing the grayscale image to be seen through the colors.

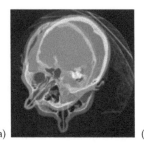 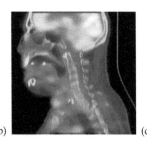 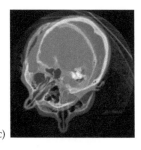

(a)                           (b)                           (c)

**FIGURE 8.17 (Please see color insert following page 16)**
Pseudocolor displays. (a) Two unregistered slices using red and green only, (b) registered PET
and CT images using red and green only, (c) two unregistered slices using red, green and blue.
(Image data from the Chapel Hill Volume Rendering Test Data Set, Volume I. [8] and http://
pubimage.hcuge.ch:8080/. With permission.)

- Image subtraction is appropriate for a pair of images from the same
  modality, and is useful for a visual check of how well the images are
  aligned. If the images had identical pixel values and were perfectly
  aligned, the result of subtraction would be zero. Differences appear
  as positive and negative values.

- In RGB pseudocolor displays, one image is displayed in shades of
  red and the other in shades of green. This means that in areas of
  overlap between the images, shades of yellow arise from the red and
  green combination. In the example in Figure 8.17a, two unregistered
  slices are displayed in this way to illustrate the principle. Note that
  the images are seen separately as red and green where there is no
  overlap, but shades of yellow appear elsewhere. Figure 8.17b shows
  a pair of registered images from different modalities with a red and
  green pseudocolor display. Another variant of the pseudocolor display uses red, green and blue, which means that areas common to
  the images appear in gray (Figure 8.17c)

- In the checkerboard display, a single image is made up of alternate
  pixels from the two registered images, in a chessboard or checkerboard layout. Providing the pixels are sufficiently small, adjacent
  pixels will be integrated by the eye, and the display looks rather like
  an overlay display. The principle is illustrated by using rather large
  blocks of pixels for the checkerboard in Figure 8.18a. In Figure 8.18b,
  the usual single pixel alternation is used, showing that it is a very
  effective display for such a simple technique.

- Several 3-D rendering and stereoscopic visualization methods are
  described in Chapter 9.

*Self-assessment question 8.08*

Match each of the words A–E, which are words associated with image registration, to the correct definition from the numbered list.

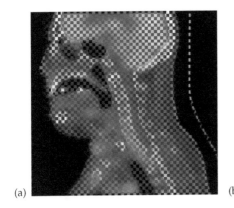 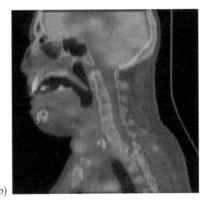

(a)        (b)

**FIGURE 8.18**
Checkerboard displays. (a) Illustration of principle using large blocks for the alternating colors of the display, (b) more realistic result where adjacent pixels are displayed in alternate colors. (Image data from http://pubimage.hcuge.ch:8080/. With permission.)

A. Translation

B. Rotation

C. Transformation

D. Corresponding points

E. Interpolation

1. A mathematical operation that might involve translation, rotation, scaling, shearing or deformation

2. Linear movement

3. Calculation of a pixel value at an intermediate position between other pixels

4. Rotation about one or more axes of the volume

5. Pairs of points sharing a property and likely to represent the same physical location in an object

## 8.5 Applications

### 8.5.1 Assessment of Disease Progression or Growth Using Temporal Series

The images may be from the same modality, but acquired at different times, for example, before and after treatment. By bringing the images into spatial alignment it is possible properly to measure any changes that have occurred. This technique is often applied to radiological image data, but the example here uses surface laser scans of the kind illustrated in Figure 8.19a. As an example, the picture in Figure 8.19b maps differences between surface laser scans taken pre- and postoperatively. Areas of the two scans that were expected to be unaffected by the change, in this case the forehead, were used to calculate the correct alignment.

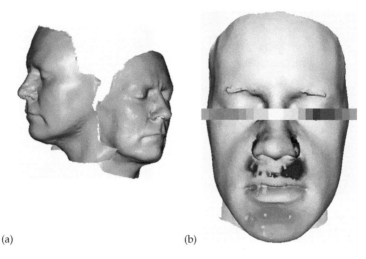

(a)                                                    (b)

**FIGURE 8.19 (Please see color insert following page 16)**
(a) Two surface laser scans acquired at different times. (b) Image calculated, for a different individual, by registering surface laser scans taken pre- and postoperatively. The color scale has 2 mm increments. The red end of the scale indicates movement backwards as a result of surgery, and the purple indicates movement forwards as a result of surgery. (From Miller, L., Morris, D.O., and Berry, E. Visualizing three-dimensional facial soft tissue changes following orthognathic surgery, *Eur. J. Orthod.*, 29, 24, 2007. With permission of Oxford University Press.)

### 8.5.2 Combination of Structural and Functional Information from Different Modalities

The images may be from different modalities, but they should be acquired at close to the same time so that it is a valid assumption that the imaged object was the same for both modalities. Bringing such data into spatial alignment can be useful if one modality gives strong anatomical information and the other holds functional information with little anatomical detail (Figure 8.17b). The two modalities need not both be large-scale radiological images. A challenging registration problem is to align microscope slides prepared for histopathological examination with, for example, magnetic resonance images. An example is shown in Figure 8.20. This can be a very difficult problem as the histology samples tend to be distorted by the physical processing that is undertaken.

A developing area for the use of image registration is in radiotherapy. In radiotherapy treatment planning, both areas to be treated and areas to be avoided are defined, and a plan devised that maximizes the radiation dose to the treatment area while minimizing the dose to radiosensitive organs. Information for defining the areas is likely to be derived from more than one imaging modality. For example, MRI has high soft tissue contrast which helps when defining the volume to be treated, but CT provides information that is required for determining the dose. Image registration is necessary so that the information from the two modalities may be combined. The need for computational speed in clinical practice, however, means that registration

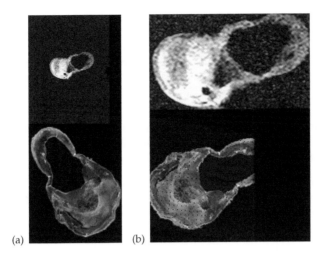

(a)                    (b)

**FIGURE 8.20**
Registration of microscope slides prepared for histopathological examination with magnetic resonance images. The upper image in each case is a T1-weighted MR image of an in-vitro carotid artery. The lower image is the corresponding histology image. (a) Before registration, (b) after registration (Courtesy of John Biglands. With permission.)

methods are specially adapted. Although areas treated in radiotherapy often involve soft tissue structures whose shape may change owing to organ motion or bladder filling, rigid registration methods applied to bone are preferred. Sometimes a preprocessing step is applied to segment the bone and further reduce the number of voxels involved in the image processing.

### 8.5.3 Creation of Atlases or Templates Representing the Typical Appearance in Health or Disease

An atlas built from medical images is a representative data set. It may combine information from several sources, or be an average of many examples that also includes information about the variability between the examples. For example, a probabilistic atlas may be used to represent a typical normal or diseased subject. By bringing images of a real subject into spatial alignment with the atlas, by image registration, it is possible to make consistent comparisons.

#### More about Anatomical Atlases

The VOXELMan project at the University of Hamburg includes several anatomical atlases; these may be viewed from links at www.uke.de/voxel-man. An early example of a brain atlas was the Talairach Atlas, which defined regions of the brain on photographs from a single subject. Talairach coordinates are used to define locations in the brain. A more recent brain atlas is the MNI brain from the Montreal Neurological Institute, which has been adopted as the standard template by the International

Consortium for Brain Mapping (ICBM) and for software used to analyze functional brain imaging data. The current standard is the average of 152 normal MRI scans, which were all matched to an earlier standard to ensure alignment. Several navigable examples can be seen at www.bic.mni.mcgill. ca/cgi/icbm_view/.

### 8.5.4    Preparation for Arithmetical or Statistical Operations, Such as Subtraction in Contrast Enhanced Data or Statistical Parametric Mapping

Analysis of contrast enhanced data does not make sense unless the images with and without the contrast agent are spatially registered, since the calculation aims to determine the change in gray value caused by the contrast agent at a particular location in the body.

### 8.5.5    Generation of a Roadmap for Invasive Procedures and Image-Guided Surgery

Registration of images to the matching physical location in the patient is necessary for image guided procedures to take place. This is currently a very active area of research.

### 8.5.6    But Image Registration May Not Be Necessary ...

Be aware, however, that it is not always necessary to perform image registration to compare information from two or more images. This is the case if it is possible to define a set of measurements to be made separately in each image, with the location of the measurement defined in terms of a relation to a feature seen in each image. For example, this might be expressed as "measure the width of X at a level midway between A and B," or "find the mean pixel value in a prescribed region of interest located at the center of C." Similarly, the volume of a tumor could be determined in two image sets without any registration, and a volume change calculated. Registration would only be required if one wanted to determine the locations at which the changes had taken place.

*Self-assessment question 8.09*

One of the following examples of image analysis tasks does not require image registration. Which one?

- Plotting the change in MR signal with time following contrast agent administration
- Determination of the change in diameter of the femoral head on serial x-ray images
- Mapping regional motion in the muscle of the heart
- Determining the pattern of movement following facial surgery

## 8.6   Good Practice Considerations

Always retain the original, acquired image data and ensure that information about how any registered image volume was generated is kept with the data.

In common with other image processing operations, when used as part of a quantitative protocol, it is necessary to justify, record and report the choices for parameters involved with image registration. These parameters are mainly associated with segmentation.

Be aware that methods of visualization that use red and green may not be very informative for those with red-green color vision deficiencies. This issue is covered in Chapter 10.

## 8.7   Chapter Summary

This chapter was concerned with the topic of image registration and the visualization of the results of registration. Registration was described in general terms as being a four-step process, and it was shown how specific implementations of registration fit into the four-step description. Definitions were provided for several commonly used terms. The reader used an ImageJ plugin to perform 2-D image registration using landmarks. Examples of typical applications requiring image registration were presented, together with guidance on how to determine whether or not registration is needed in an image analysis protocol.

## 8.8   Feedback on Self-Assessment Questions

### Self-assessment question 8.01

The laser surface scanner is useful for measuring changes in surface shape, caused, for example, by swelling or as a result of surgery. Similar systems are used for positioning purposes during computer-assisted surgery or oncology, or by dentists making replacement teeth. Surface data are also valuable for making close fitting masks for treating burns or for restraining a body part during treatment.

### Self-assessment question 8.02

"A surface matching technique has been developed to register multiple imaging scans of the brain in three dimensions, with accuracy on the order of the image pixel sizes."[3] The paired features in this case are points on a surface.

"This method does not require fiducial markers and the user is not required to identify common structures on the two image sets. To align the images, the algorithm seeks to minimize the standard deviation of the PET pixel values that correspond to each MRI pixel value."[4] The paired features in this case are pixel values.

"The technique is based on user identification of point-like landmarks visible in both modalities."[5] The paired features in this case are anatomical landmarks.

### Self-assessment question 8.03

Figure 8.10b represents a shearing deformation and Figure 8.10c is a rotation of the shape in Figure 8.10a.

### Self-assessment question 8.04

The transformation described in the extract [6] is an affine transform, because it involves a rescaling transformation. If only translation and rotation had been used, it would have been a rigid transformation.

### Self-assessment question 8.05

Figure 8.11b shows an affine transformation, Figure 8.11c shows a projective transformation, Figure 8.11d shows a rigid transformation, Figure 8.11e shows a curved transformation.

### Self-assessment question 8.06

The expected gray value resulting from subtraction of a pair of identical images is (c) zero. Of course, real registered images won't give a perfect result, but if the mean gray level is not approximately zero, it is worth checking out. Note that some software does not use signed integers, and a constant is automatically added to ensure that all the pixel values in the result are greater than zero. You can check that the operator is acting as you think by subtracting an image from an identical copy of the image.

### Self-assessment question 8.07

Assessment of disease progression and monitoring patient response to treatment are both single modality examples. Determination of the anatomical location of functional data requires at least two modalities, one showing functional information and one showing anatomical information, so it is a multimodality registration task.

## Self-assessment question 8.08

A. Translation, (2) Linear movement

B. Rotation, (4) Rotation about one or more axes of the volume

C. Transformation, (1) A mathematical operation that might involve translation, rotation, scaling, shearing or deformation

D. Corresponding points, (5) Pairs of points sharing a property and likely to represent the same physical location in an object

E. Interpolation, (3) Calculation of a pixel value at an intermediate position between other pixels

## Self-assessment question 8.09

Image registration is not required to determine the change in diameter of the femoral head on serial x-ray images. In this option, the measurement required is a diameter, which one should be able to measure reliably from each of the images separately with no need for registration. In the other examples, there is a need to link changes to a set of particular spatial locations, and image registration is required.

## References

1. The BrainWeb Simulated Brain Database. www.bic.mni.mcgill.ca/brainweb/.
2. Collins, D.L. et al. Design and construction of a realistic digital brain phantom, *IEEE Trans. Med. Imag.*, 17, 463, 1998.
3. Pelizzari, C.A. et al. Accurate 3-dimensional registration of CT, PET, and or MR images of the brain, *J. Comput. Assist. Tomogr.*, 13, 20, 1989.
4. Woods, R.P., Mazziotta, J.C., and Cherry, S.R. MRI-PET registration with automated algorithm, *J. Comput. Assist. Tomogr.*, 17, 536, 1993.
5. Hill, D.L.G. et al. Registration of MR and CT images for skull base surgery using point-like anatomical features, *Br. J. Radiol.*, 64, 1030, 1991.
6. Holman, B.L. et al. Computer-assisted superimposition of magnetic resonance and high-resolution technetium-99m-HMPAO and thallium-201 SPECT images of the brain, *J. Nucl. Med.*, 32, 1478, 1991.
7. Thévenaz, P., Ruttimann, U.E., and Unser M. A pyramid approach to subpixel registration based on intensity, *IEEE Trans. Image Proc.*, 7, 27, 1998.
8. The Chapel Hill Volume Rendering Test Data Set, Volume I, Department of Computer Science, University of North Carolina. Available at http://education.siggraph.org/resources/cgsource/instructional-materials/volume-visualization-data-sets.

# 9

## Visualization and 3-D Methods

### 9.1 Introduction

Visualization is the display of image data. The display need not be a complicated presentation of data, and any image display can be described as a visualization, be it two or three dimensional. However, medical image visualization does go beyond viewing a piece of film on a light box, as it provides a way of dealing with the vast number of cross-sectional images acquired in routine practice. Specialized approaches are needed because objects visualized have a huge range of scales, ranging from molecules and cells to whole body parts. Furthermore, different attributes of these objects beyond their anatomical appearance can be visualized, including biophysical, biomechanical and physiological properties. The results of visualization may be used for diagnosis, treatment planning, rehearsal, assessment and intraoperative guidance.

#### 9.1.1 Learning Objectives

When you have completed this chapter, you should be able to

- Understand the advantages and disadvantages of different visualization methods
- Define scene-based and object-based visualization methods
- Give examples of alternative ways of viewing data as slices
- Discuss 3-D rendering
- Compare and contrast maximum intensity projection (MIP) with surface rendering
- Give examples of the cues used to give an impression of three dimensionality
- Use ImageJ to perform 3-D rendering

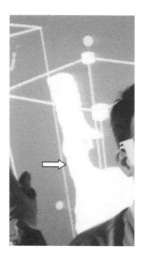

**FIGURE 9.1**
Image for self-assessment question 9.01, a 3-D rendering generated from a volume of x-ray CT data is arrowed. (Courtesy of John Truscott. With permission.)

### 9.1.2   Example of an Image Used in This Chapter

*Self-assessment question 9.01*

Figure 9.1 shows part of a figure that appears later in the chapter. It shows a 3-D rendering (arrowed) generated from a volume of CT data. A CT slice that was part of the data volume appears elsewhere in this book. In which chapter does that image appear?

## 9.2   Taxonomy for Visualization

Visualizations may be divided into two main groups, which form part of a taxonomy [1] that helps make the range of techniques for visualization more manageable (Figure 9.2):

- In *Scene-based* visualization, all the image information is available for rendering
- In *Object-based* visualization, explicit definition of objects (e.g., by segmentation of a surface) is performed before rendering

## 9.3   Scene-Based Visualization

Within the scene-based category, there are again two categories, slice mode and volume mode.

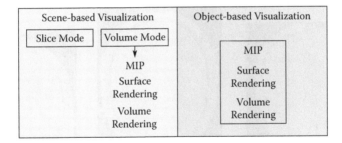

| Scene-based Visualization | Object-based Visualization |
|---|---|
| Slice Mode \| Volume Mode | MIP |
| MIP | Surface Rendering |
| Surface Rendering | Volume Rendering |
| Volume Rendering | |

**FIGURE 9.2**
Hierarchical classification of visualization methods. (Adapted from Udupa, J.K. 3D Imaging: Principles and approaches, in Udupa, J.K. and Herman, G.T., Eds., *3D Imaging in Medicine*, 2nd ed., CRC Press, Boca Raton, 2000. With permission.)

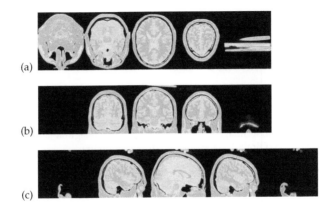

(a)

(b)

(c)

**FIGURE 9.3**
Orthogonally resliced data. (a) Five slices from the acquired set of transverse slices, (b) five coronal slices calculated from transverse stack, (c) five sagittal slices calculated from transverse stack. (Image data from the BrainWeb Simulated Brain Database [2,3]. With permission.)

### 9.3.1 Slice Mode

A 2-D view in the slice mode may be a projection (a conventional, plain x-ray) or a representation of a slice (MR or CT slice). The simplest example of slice mode visualization is to view an acquired slice. To aid interpretation, that slice can have gray-level windowing applied or be viewed using a color look-up table. For multislice sets, the images can be displayed one at a time in a window that allows the user to move through the image stack or as a montage.

There is, however, no restriction that the original slices must be viewed. The 3-D information can be used to generate new 2-D views. For example, a stack of transverse slices may be sliced in another direction. Figure 9.3 shows selected slices from the original stack of transverse slices, together with coronal and sagittal slices calculated from that data. This kind of capability is interesting, for example, in ultrasound imaging, where limited acquisition

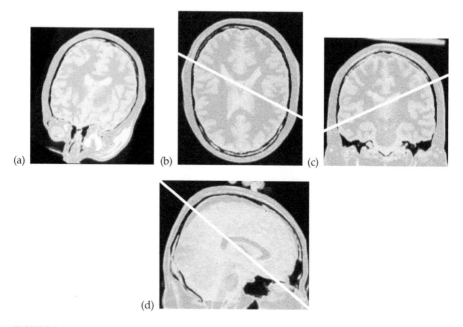

**FIGURE 9.4**

(a) Oblique slice calculated from the acquired set of transverse slices, (b) navigation image showing slice location on transverse slice, (c) navigation image showing slice location on coronal slice, (d) navigation image showing slice location on sagittal slice. (Image data from the BrainWeb Simulated Brain Database [2,3]. With permission.)

windows prevent the direct acquisition of certain views. Instead, a view may be calculated from the acquired data, though of course if particular features were not visible on the original slices, they will not be seen on the reformatted data either. This approach is known as *reslicing* or *multiplanar sectioning and display*. The calculated slices can be orthogonal (transverse, sagittal and coronal) as in Figure 9.3, oblique (i.e., at an angle other than 90° to the acquired slices) as shown in Figure 9.4, or even show curved surfaces (possibly following the shape of an anatomical structure). The curved surface option is harder than the others to envisage. The principle is illustrated in Figure 9.5. A line is drawn on one slice in the stack (Figure 9.5a), and this is reproduced through the stack of slices (Figure 9.5b) to make a curved surface (Figure 9.5c). An image is created where all data lying on this newly defined surface are mapped onto a flat image (Figure 9.5d). The use of curved slices is a rather specialized method that tends to be used by observers who already have a good mental picture of the form of the 3-D object.

## Activity: Slice Mode Visualization in ImageJ

Load the image stack chapter09.tif* from the CD. A window appears with a scroll bar across the bottom. The scroll bar may be used to move through

---

* Image data from the BrainWeb Simulated Brain Database [2, 3]. With permission.

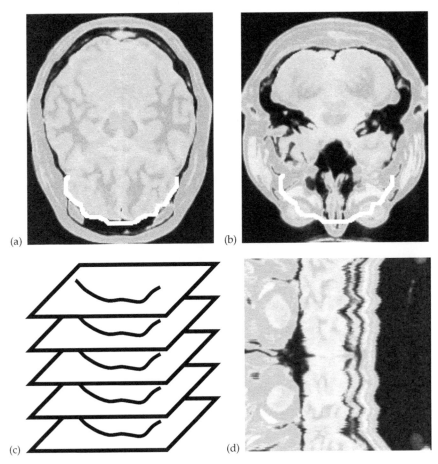

**FIGURE 9.5**
Generation of images of a curved surface. (a) The white line represents the location of curved surface. The line is drawn on one transverse slice. (b) Example of the drawn white line reproduced on another image in the stack. (c) The line is reproduced throughout the stack to represent a curved surface. (d) Image showing data on the defined curved surface. (Image data from the BrainWeb Simulated Brain Database [2,3]. With permission.)

the stack of images; the image number is displayed in the top left of the window.

*Self-assessment question 9.02*

To display the images as a montage, there is an option in the Image | Stacks menu of ImageJ. Find the right option and choose values for the parameters that will give a montage like the one shown in Figure 9.6.

## Activity: Slice Mode Rendering Using ImageJ

Use the same stack as before (chapter09.tif).

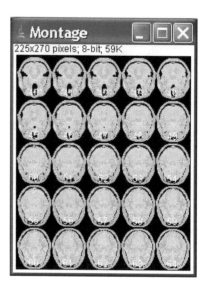

**FIGURE 9.6**
Image montage for self-assessment question 9.02

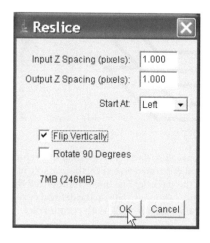

**FIGURE 9.7**
The ImageJ Reslice dialog.

Select Image | Stacks | Reslice[/]... and enter the settings shown in Figure 9.7.

Scroll through the new slices and find one that shows a clear section through one of the eyes. Compare your selection with Figure 9.8. If the head appears upside down, then perhaps the tick beside Flip Vertically in the Reslice dialog box was omitted.

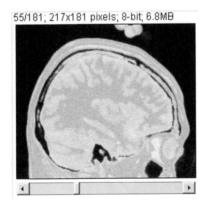

55/181; 217x181 pixels; 8-bit; 6.8MB

**FIGURE 9.8**
Result of slice mode rendering activity. (Image data from the BrainWeb Simulated Brain Database [2,3]. With permission.)

*Self-assessment question 9.03*
In the stack used in the activities, the voxels are cubic, that is, they are the same size in all three directions, and the slice thickness is the same as the pixel size in each transverse slice. Often, in practice, thicker slices are acquired. Describe how the resliced image would differ if the slice thickness were two times the in-slice pixel size.

### 9.3.2 Volume Mode

The volume mode category is useful for visualizing 3-D data that contain surfaces, interfaces or intensity distributions. There are three subcategories:

- Maximum intensity projection (MIP)
- Surface rendering
- Volume rendering

In all cases, projection is required to get from the three-dimensional data to a two-dimensional rendering on a viewing plane. Projection is accomplished by a computational technique called *ray tracing* or *ray casting* (Figure 9.9). This involves either tracking from each pixel in the viewing plane through the volume of data, or performing element projection (sometimes called splatting) to project voxels onto the viewing plane. The tracing may be done either using parallel rays or with perspective. Perspective will render closer parts of volume larger in size. The process is repeated for a sequence of viewing planes to give the impression that the object is rotating in front of the viewer. The information about each voxel that is projected along the ray depends on the type of rendering being performed.

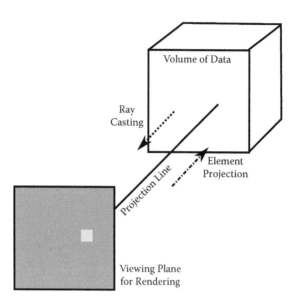

**FIGURE 9.9**
Projection in volume mode rendering.

### 9.3.2.1 Maximum Intensity Projection (MIP)

Maximum intensity projection is achieved by projecting only the brightest voxel that occurs along the projection line. It works well when the object of interest is the brightest feature in the image and where the image information is quite sparse (when there is a small number of very bright features). MIP is widely used, especially for vascular applications such as contrast-enhanced CT and magnetic resonance angiography (MRA).

Figure 9.10 shows a single MRA image taken from a stack of slices. This kind of image is an example that would be good for MIP; four blood vessels are the brightest features in an image that has little other content. Further slices from the stack are shown in Figure 9.11, and MIP renderings from three different viewpoints are in Figure 9.12. MIPs display best as movies, as these give the impression of three dimensions and the relative positions of vessels can be seen.

### 9.3.2.2 Surface Rendering

For surface rendering, the surface to be rendered is defined, usually by gray-level thresholding, and cues are added to the rendering to give it a 3-D appearance. The cues for 3-D appearance include:

- Hidden surface removal to mimic how things appear in real life. If another object is blocking the view of an object, then it is not seen.

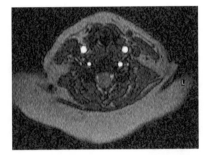

**FIGURE 9.10**

A magnetic resonance angiography (MRA) image. A stack of such slices (Figure 9.11) is suitable for rendering by maximum intensity projection (MIP). (Courtesy of John Ridgway. With permission.)

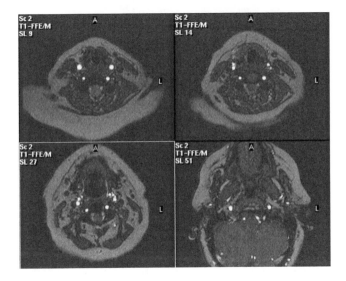

**FIGURE 9.11**

Selected slices from a stack of magnetic resonance angiography (MRA) images. (Courtesy of John Ridgway. With permission.)

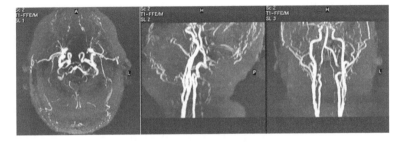

**FIGURE 9.12**

Maximum intensity projection renderings from three different viewpoints, all calculated from the data of Figure 9.11. (Courtesy of John Ridgway. With permission.)

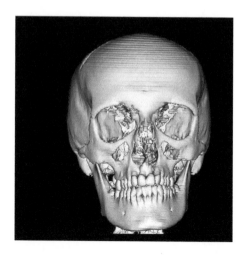

**FIGURE 9.13**

Surface rendering of a skull with shading to improve the 3-D appearance of the surface shape. (Image data from the Chapel Hill Volume Rendering Test Data Set, Volume I. [4] With permission.)

- Shading is used to help give the impression of surface shape and angle. Thus the sides of an object appear darker than the surface perpendicular to the viewer. An example of shading is shown in Figure 9.13.

- Stereoscopic displays mimic how the observer normally perceives three-dimensional objects. Each eye receives a slightly different view of the object, and these views are combined in the brain to give the impression of depth. In visualization, this is accomplished by presenting the observer with two slightly different views of the object, one as would be seen by the left eye, and one as seen by the right. A range of methods can then be used to ensure that the left-eye view is seen only by the left eye and the right-eye view only by the right eye.

  - The pair of images may be colored so that when viewed using colored filters only the correct image is seen by each eye. The colors red and green, or red and cyan are commonly used. The images are called *anaglyphs* or *stereograms*. The filters are held in cardboard frames to make a simple pair of glasses. This is the method used to present 3-D movies in the cinema.

  - An alternative is glasses with built-in shutters that open and close very rapidly and allow the viewer to see through one lens and then through the other (Figure 9.14). The images on the computer screen are changed between the left- and right-eye views at the same rate. The persistence effect in the brain means that the images are blended into one.

**FIGURE 9.14**
Goggles with built-in shutters worn to give 3-D perception in a virtual environment. (Courtesy of John Truscott. With permission.)

- The impression of three-dimensionality is reinforced by making the object appear to rotate, so that the shading changes as the object moves relative to the light source. A movie of this kind is the output of the visualization case study in Chapter 11.

- Cast shadows also help give an impression of three-dimensionality.

- In the field of computer science, texture is data of any kind that is mapped onto a surface. So in addition to the obvious textures such as rough and smooth surfaces, a color image would also be described as texture when mapped onto a surface. For 3-D rendering, the texture is often designed to make the surface look shiny, so that the 3-D appearance is enhanced (Figure 9.15).

### 9.3.2.3 *Volume Rendering*

Volume rendering is very similar to surface rendering, but instead of defining the object surface in absolute terms (so that voxels are either on the surface, or they are not) a "fuzzy" object is used. In this case, each voxel has a value that indicates how much it resembles a voxel that is on the surface of interest. In surface rendering, these values can only be 0% or 100%, but in volume rendering a voxel could be defined as being, for example, 30% like the surface. For rendering purposes, each voxel is assigned an opacity that depends on how much that voxel resembles the desired surface, and that opacity determines the transmission, emission and reflection from that voxel when the rendering is performed. The fuzzy approach is very useful

**FIGURE 9.15**
Surface rendering of a vertebra using a shiny texture map to enhance the perception of 3-D shape.

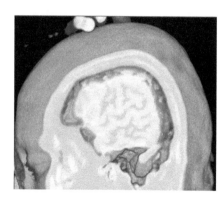

**FIGURE 9.16**
A volume rendering generated from the stack of images shown in Figure 9.3. (Image data from the BrainWeb Simulated Brain Database [2,3]. With permission. VolumeViewer authored by Kai Uwe Barthel from http://rsb.info.nih.gov/ij/plugins/volume-viewer.html. With permission.)

for medical imaging, where partially volumed voxels are common. The *partial volume effect* arises because each voxel may represent more than one tissue type, and so have a value that does not clearly belong to one tissue or another. The use of a fuzzy approach means that there is no need to make a decision to include or exclude a voxel. Each voxel will be included, but with a fuzzy value that will make its contribution relatively weaker if necessary.

The other feature associated with volume rendering that distinguishes it from surface rendering is that the full set of data remains available, meaning that rather complex visualizations including surfaces, slices and objects can be generated. Figure 9.16 is a volume rendering calculated from the same stack of images as in Figure 9.3. It is clear that the full set of data was available for the rendering, since both the surface and detail of an interior slice are displayed.

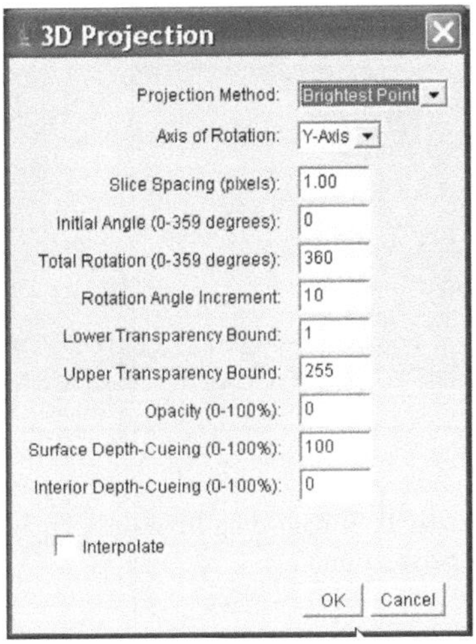

**FIGURE 9.17**
The ImageJ 3-D Projection dialog.

## Activity: 3-D Rendering Using ImageJ

Load CTheadsagy.tif* into ImageJ from the CD. This has 256 slices, so may take time to load on slower computers. Scroll through the images in the stack. Which parts of the data would you expect to see emphasized in a maximum intensity projection? Make a note of your thoughts.

Select Image | Stacks | 3-D project.

Set the parameters shown in Figure 9.17, then click OK. Wait for all 36 frames in the sequence to be calculated (Figure 9.18).

To play the movie that has just been created, select Image | Stacks | Start Animation, and to stop the movie select Image | Stacks | Stop Animation.

Did you correctly predict which part of the data would be emphasized in the maximum intensity projection, the tooth fillings and bone?

In the previous example, the reconstructions were made for views at various angles around the y axis. Think about which way you would expect the head to rotate if projections about the x axis were requested, then try it and see if you were right. You should find that the y axis is vertical, x is horizontal and the z axis goes into the monitor.

---

\* Image data from the Chapel Hill Volume Rendering Test Data Set, Volume I. [4] With permission.

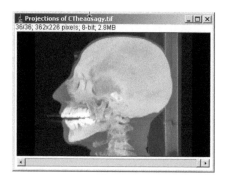

**FIGURE 9.18**
Result of the 3-D rendering activity. (Image data from the Chapel Hill Volume Rendering Test Data Set, Volume I. [4] With permission.)

## 9.4   Object-Based Visualization

MIP, surface rendering and volume rendering can also be performed under the object-based category in the taxonomy. The difference from scene-based visualization is simply that for object-based visualization the image data have been preprocessed to include data only from the object or objects of interest. Methods of segmentation are covered in Chapter 2.

*Self-assessment question 9.04*

Which of the following statements are true?

   A. 3-D glasses work by ensuring that each eye sees an image that is offset by about 45° from the image seen by the other eye.
   B. 3-D glasses work by ensuring that each eye sees an image that is offset by about 5° from the image seen by the other eye.
   C. 3-D glasses rely on the persistence of vision effect in the brain.
   D. If shutter 3-D glasses are used, it is necessary for the left- and right-eye images to be colored in cyan and red.

*Self-assessment question 9.05*

Which of the following statements are true?

   A. A movie that shows the 3-D object zooming towards and away from the viewer improves the perception of 3-D shape.
   B. A movie that shows the 3-D object rotating improves the perception of 3-D shape.
   C. The use of warm and cold colors improves the perception of 3-D shape.

D. The use of shading and shadows improves the perception of 3-D shape.

E. Montages improve the perception of 3-D shape.

## 9.5   Other Visualization Methods

### 9.5.1   Parametric Displays

Parametric displays are another aid to the visualization of complex data sets. In a parametric display, further information is mapped onto a 2-D image or a 3-D surface, usually displaying the value of a parameter associated with that location. Figure 9.19 shows a pair of terahertz-frequency images of a thin slice of a tooth. Each image shows values of a different parameter associated with the variation of attenuation with frequency in the tooth slice. In an example from the previous chapter, in Figure 8.19b, information has been combined to show changes that have occurred because of jaw surgery. Colors to indicate the amount of change have been mapped onto the postoperative scan, and show that the upper and lower jaws have been moved in opposite directions.

### 9.5.2   Virtual Environments

Virtual reality systems and virtual environments not only provide a visualization of the data, but add the ability to interact with a rendered scene. In medicine, they have uses analogous to flight simulators, and are attractive for training purposes and for the rehearsal of complex procedures. In the virtual environment in Figure 9.14, data gloves are worn to allow the position and orientation of the hand to be tracked. Additionally, force feedback is used to make the feel of an object or tool realistic. Virtual environments rely heavily

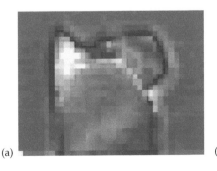 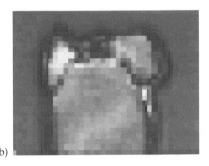

(a)                                    (b)

**FIGURE 9.19**
Parametric terahertz pulsed imaging images of a 250-μm slice of a tooth. In each case the imaged parameter is taken from a graph of attenuation coefficient plotted against frequency (a) slope and (b) intercept. The pixel size is 0.33 mm.

on image registration methods (Chapter 8) as it is necessary to perform tracking and to register several different components to a common coordinate system (e.g., images, computer modeled tools, real tools and operator).

### 9.5.3 Physical and Hybrid Representations

In the engineering technique of rapid prototyping, image data are used to build a physical model, instead of being visualized on the computer. For some applications, a model may be more appropriate than visualization, especially those where the user wants to pick up and feel the object. Rapid prototyping has also found a role in the manufacture of patient-specific tools for training and for use during surgery. Figure 9.20 shows examples of tubular, flexible silicone rubber models that fit into a training simulator for the keyhole repair of abdominal aortic aneurysms. Figure 9.21 shows a drilling template for pedicle screw insertion. The template fits uniquely in place on an individual's spine, having been designed using preoperative CT data. The CT data were also used to plan the best locations for the screws, and guide-holes were built into the template to ensure that, intraoperatively, the holes are made in the right place.

## 9.6  Good Practice Considerations

Information must be recorded and reported about two aspects of the visualization process:

- How the surface of the rendering was defined (e.g., segmentation details for object-based renderings)
- Details of the rendering method and the values of parameters chosen, as these can affect the appearance

The original slices from which a rendering has been made should be available for review by an observer if requested.

Be aware of strategies to preserve anonymity when facial data are part of an image volume.

## 9.7  Chapter Summary

In this chapter, a classification system was used to differentiate the methods for 2-D and 3-D image visualization. A distinction was made between

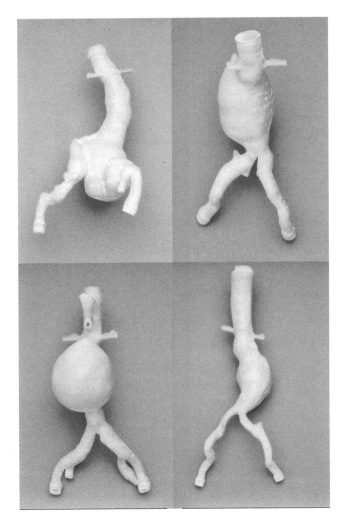

**FIGURE 9.20**
Examples of tubular, flexible silicone rubber replicas generated from x-ray CT data. They represent typical anatomical variations of human abdominal aortic aneurysms and were made as part of a training simulator. (From Berry, E. et al. Proc. Inst. Mech. Engineers Part H: *J Eng. Med.*, 216, 211, 2002. Permission is granted by the Council of the Institution of Mechanical Engineers.)

visualizations for which all the image data are available to be included in the rendering and those where a specific object has already been extracted from the data. Definitions for maximum intensity projection, surface and volume rendering were given. Further methods, including physical renderings and virtual environments, were also briefly described. The reader performed 3-D rendering using ImageJ.

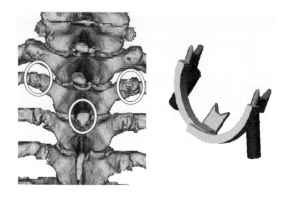

**FIGURE 9.21**
A drilling template designed from x-ray CT data for pedicle screw insertion in the lumbar region. (From Berry, E. et al. Proc. Inst. Mech. Engineers Part H: *J Eng. Med.*, 219, 111, 2005. Permission is granted by the Council of the Institution of Mechanical Engineers.)

## 9.8   Feedback on Self-Assessment Questions

### Self-assessment question 9.01

Chapter 2. The CT slice, which was part of the volume of CT data from which the 3-D rendering was generated, appears in several figures including Figure 2.1. Note that the replica shown at the bottom left of Figure 9.20 was generated from the same volume of CT data. The images show a large abdominal aortic aneurysm.

### Self-assessment question 9.02

The menu option to display the images as a montage is Make Montage … The settings used are shown in Figure 9.22.

### Self-assessment question 9.03

If the slice thickness were two times the in-slice pixel size, the sagittal slices would have rectangular pixels and not square pixels. The rectangular pixels would be larger in the direction from head to toe. The in-slice resolution of the resliced images will be worse than for cubic voxels. If the reslicing program retained the settings associated with cubic voxels, then the sagittal slices would appear to be squashed in the head-to-toe direction (Figure 9.23a). Alternatively, if the correct voxel size was supplied to the program, then the geometry of the reslice would be correct, but the rectangular pixels give poorer spatial resolution (Figure 9.23b) than in the original (Figure 9.8). Interpolation is often used to ensure that the voxels in a stack of slices are isotropic, with the same size in all three dimensions. The simplest methods

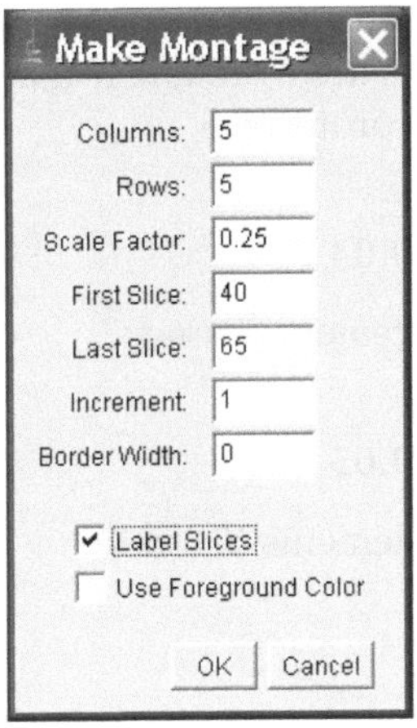

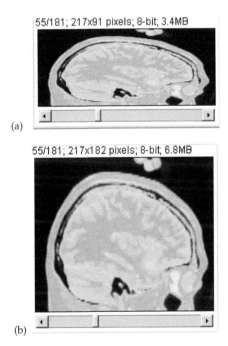

**FIGURE 9.22**
Feedback for self-assessment question 9.02, showing the settings used in the ImageJ Make Montage dialog.

**FIGURE 9.23**
Images for feedback on self-assessment question 9.03. (a) Reslice with slice thickness set to half its correct value, (b) Reslice with correct slice thickness, which is twice the size of the in-plane voxel dimension. (Image data from the BrainWeb Simulated Brain Database [2,3]. With permission.)

of interpolation (Chapter 1) work well for original size ratios up to about 5, and more complex methods are available if this limit cannot be avoided by using a revised acquisition protocol.

**Self-assessment question 9.04**

The true statements are statements B and C.

**Self-assessment question 9.05**

The true statements are statements B and D.

## References

1. Udupa, J.K. 3-D Imaging: Principles and approaches, in *3-D Imaging in Medicine*, Udupa, J.K., and Herman, G.T., Eds., CRC Press, Boca Raton, 2000, chap. 1.
2. The BrainWeb Simulated Brain Database. www.bic.mni.mcgill.ca/brainweb/.
3. Collins, D.L. et al. Design and construction of a realistic digital brain phantom, *IEEE Trans. Med. Imag.*, 17, 463, 1998.
4. The Chapel Hill Volume Rendering Test Data Set, Volume I, Department of Computer Science, University of North Carolina. Available at http://education.siggraph.org/resources/cgsource/instructional-materials/volume-visualization-data-sets.

# 10

## Good Practice

## 10.1  Introduction

Each of the preceding chapters has included a section on good practice, high-lighting points particularly relevant to the topic of the chapter. It is hoped that those experienced in the field will have found these points obvious, as most researchers and practitioners absorb the principles of good practice as they gain experience. However, for those starting out, some issues may not be clear. In the earlier chapters, points were made that were related to the relevant image processing techniques. In this chapter the same points are grouped into categories that cut across the classes of image processing: scientific rigor, ethical practice, human factors and information technology. An important aspect of medical image processing is evaluation of newly developed image analysis protocols. An introduction to evaluation of segmentation, registration and visualization algorithms is given.

### 10.1.1  Learning Objectives

When you have completed this chapter, you should be able to

- Give examples of information that should be included in a description of an image analysis protocol.
- Recognize, and avoid, common pitfalls associated with medical image processing.
- Describe qualitative and quantitative methods for evaluating image processing results.
- Explain why reproducibility should be assessed using different images from the same subject.

## 10.2  Scientific Rigor

The availability of software, intended for the home user and able to perform a range of image processing operations, can encourage an ad hoc attitude to image processing. The user will try a number of options, and see which one makes the image look good. Such an ad hoc approach to medical image processing is, however, inappropriate. In spite of its obvious visual appeal, medical image processing is a quantitative field in which a subjective approach must be avoided. Medical image processing, whether being applied in clinical practice or used in a research study, is a scientific topic and so the aim is to obtain and report the results in the usual scientific way. The reader must be given sufficient information to repeat the work, and this requirement for complete reporting is the principle underlying many of the good practice points included in this chapter.

### 10.2.1   Protocols and Workflows

It is important to ensure that a clear protocol, both for acquiring the original images and performing the analysis, is devised and followed. Such protocols are also known as image data workflows, to emphasize their relationship to workflows used in other applications. A workflow is a defined sequence of steps that is required to accomplish a particular task. It involves the coordination of people and technology, and includes full documentation to provide an audit trail of activity.

Consistency in the acquisition protocol is required so that it is valid to combine or compare results from data acquired at different times, and so that exactly the same analytical procedure may be used on the data. For example, variations in spatial or grayscale resolution from subject to subject, or from acquisition to acquisition, will compromise results obtained by segmentation or classification methods. Acquisition systems often have automated setup routines, which can be helpful in generating images for human review, but they may introduce undesired differences between subjects when later image analysis is planned. It is also necessary to pay attention to acquisition-related operations such as image storage and transfer, to ensure that, for example, lossy compression is not inadvertently included by the default settings of a system. The preceding points underline the need for a good dialog to exist between those acquiring the images and those who will perform the analysis (assuming they are not the same individual). When good communication is in place, the analyst will understand the restrictions on acquisition, and the imager will recognize the reasons for what may seem to be overcautious requests regarding acquisition and storage.

Consistency is also essential in image analysis, and a written protocol will make clear the sequence of operations to be undertaken. Advice will be included when steps are required to improve consistency, or if instructions for interactive image processing might be open to interpretation. Such advice might include a standard procedure for setting up the monitor display, warnings to avoid extra processing steps such as edge enhancement, and a definition of whether to place a line for segmentation on object pixels or on pixels outside the object.

## 10.2.2 Avoid the Arbitrary or Subjective

A good practice point that was emphasized in several chapters is the need to avoid arbitrarily or subjectively chosen values for parameters used in image processing operations. Examples include the choice of gray value used for thresholding or region definition, the size and shape of the structuring element in morphology, the choice of filter for spatial and frequency domain operations, and decisions on whether to use a global or local operation. There will be two parts to the justification. Firstly, consider the relevance of the chosen parameter to the spatial and grayscale properties of the image feature to be removed or enhanced. It is rare that the values previously reported in a published article can be used, because variations in imaging systems mean that the image being analyzed will differ in key properties such as spatial resolution. Secondly, steps should be taken to ensure that the choice of parameters is equally valid for all the images processed within a study. Where there is anatomical variation between subjects, it may be necessary to set size parameters as a proportion of the size of a larger anatomical feature, and not in absolute terms.

## 10.2.3 Recording and Reporting

The reasons for the choice of a method and values chosen for parameters must always be fully reported. This is a requirement of scientific reporting rather than an issue specific to image processing. An aspect that may be overlooked in this respect is associated with the use of image enhancement to produce cosmetic changes to make an image more informative for publication. Although these changes are not strictly part of the scientific protocol, they are potentially misleading for the reader trying to reproduce the reported work. For example, consider a report of a modified method of image acquisition, where the image contrast in published images was enhanced for visual effect. A reader might be frustrated in efforts to reproduce the contrast properties in a published image, and not realize that they had been achieved by post processing. Of course, the reader too must take responsibility and ensure that all factors associated with imaging have been properly understood. If the reader fails to note, for example, that anesthetized or nonliving subjects have been used, then replication of results will be difficult because higher image quality can achieved with stationary subjects.

## 10.2.4 Data Integrity and Image Provenance

A key good practice point is the need always to keep the original image data safe, and this is the first step in maintaining data integrity. Furthermore, care must be taken regarding image formats, with only lossless compression applied. It is wise to save processed images frequently and with different, and informative, file names. In a large-scale study, the image data workflow should include a system for naming files saved at different points in the processing. This procedure is sometimes called *versioning*. The workflow can be used to

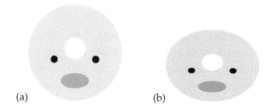

(a)                    (b)

**FIGURE 10.1**
(a) Original image. (b) The same image displayed in a box with a different aspect ratio between the width and the height.

ensure that the provenance (or history) of an image is maintained in an audit trail of activity. Such an audit trail means that it is always possible to determine information associated with a given image, including acquisition details, and a note of all the processing that has been undertaken since acquisition.

Even if great care is taken over maintaining the integrity of the data, mistakes can still be made with image display. Observers expect to see the data displayed with the correct spatial proportions, but unfortunately some software packages default to displaying image data in a fixed box size, thus losing the original aspect ratio between width and height. This effect is illustrated in Figure 10.1. Maintain the same vigilance about display when using nonscientific software tools such as word processing, poster or presentation packages. Compare the displayed images with the originals to ensure that they have not been distorted.

## 10.3   Ethical Practice

### 10.3.1   Informed Consent

It is necessary for the practitioner or researcher working with medical images to be familiar with local ethical practice. In many cases, an image is considered in the same way as a sample of tissue, and so is subject to constraints regarding its use. While the originator of a large study will be well aware of the need to obtain permission for the study and seek informed consent, a pitfall to be avoided is the casual reuse of image data in a small-scale study, perhaps for a student project, without adhering to the same constraints.

### 10.3.2   Anonymization

Names and identifying data should always be removed from image data. The increased use of 3-D data sets of the head leads to a further challenge to anonymization. Information in the image data is sufficient to reconstruct a recognizable face, and so presents a privacy problem. For some imaging applications not involving the face itself, software can be used to remove the facial information from data but retain the data required for analysis.

## 10.4 Human Factors

One of the drivers behind automation of image analysis is to exclude the subjectivity associated with the human observer. However, human observers are integral to many studies in medical imaging, and even studies using automated analysis will have included evaluation of the methods against results from human observers at some stage in their development.

### 10.4.1 Variability

Subjective image enhancement may, of course, be used to adjust the display of an image for personal preference, providing that no changes based on these settings are made to the stored data, and no further analysis is made on this personalized display. In some research studies, skilled observers are used to place images into categories, such as diseased or nondiseased. The observers will adjust the image display in a manner that they know from experience helps them to perform at their best. No numerical values are drawn from the image data, and freedom of viewing preferences is, in this situation, acceptable. When studies performed in this way are written up, information would be included regarding the observers' freedom to window the images as desired, and to adjust the ambient lighting.

Freedom to adjust display parameters is less appropriate where a manual segmentation task is to be performed. Here personalized settings could make a difference to the apparent location of an edge in the image. As indicated previously, it is usual for the study protocol to include instructions designed to reduce variability from sources like this.

Human observers tire quickly, both physically and mentally. A good study design will plan viewing sessions to avoid fatigue as much as possible. This is an important consideration, as viewers may have to view the images several times for assessment of intraobserver variability, for which each observer usually reports on the same images at least two times. The second viewing will take place after a gap of some weeks to reduce decisions based on the memory of what was done the first time.

### 10.4.2 Color Vision Deficiency and Pseudocolor Displays

The results of medical image processing are sometimes visualized using the colors red and green. This is partly because standard displays have red, green and blue color channels, and software makes it easy to send information to

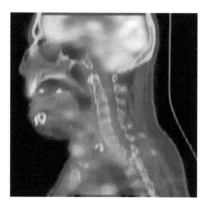

**FIGURE 10.2 (Please see color insert following page 16)**
The appearance of Figure 8.17b to a viewer with the severe red–green color deficiency deuter-anopia. (This image was generated using software from www.vischeck.com. Image data from http://pubimage.hcuge.ch:8080/. With permission.)

each channel separately. Also, the colors red and green contrast well with each other, at least they do for an observer without a color vision deficiency. However, images using just red and green for comparison purposes will be less useful to an important proportion of viewers, as 8% of males have color vision deficiency of some sort and most of these deficiencies are in red-green perception. The pseudocolor display is an example in which a pair of images is combined into one, with one of the images displayed in shades of red and the other in shades of green (Figure 8.17b). Figure 10.2 shows how the image in Figure 8.17b would appear to someone with the severe red-green color deficiency deuteranopia. There is no perceptible difference between the component images. A similar potential for confusion arises with any method that relies on red and green being perceived very differently, even something as basic as a color look-up table. Color vision deficiencies are worth bearing in mind when preparing image displays that are intended to enhance the viewer's understanding. However, one type of red-green display is not affected by color vision deficiencies, and this is discussed in the next section.

### 10.4.3 Color Vision Deficiency and Stereograms Viewed Using Red and Green Glasses

Stereogram images, also known as anaglyphs, are made up of two compo-nent images, each of which is displayed using a different color. The colors used are often red and green, and an example is shown in Figure 11.24. The two component images represent the offset views of a 3-D object seen by each eye. Spectacles with colored filters are used to view the anaglyph, so that each eye receives only the image designed for its angle of view. The viewing of anaglyphs is not prevented by red-green color blindness because the color is important only to ensure that each image passes through a dif-ferent colored filter before reaching the eye. It is not necessary for the eye to distinguish the colors, and so red-green anaglyphs can be viewed success-fully by people with a color vision deficiency. For certain color vision defi-ciencies, the stereogram image seen may be slightly suboptimal because the red image appears as a darker shade than the green image.

## 10.5   Information Technology

### 10.5.1   Software Tools

There are now many software tools available that allow the nonprogram-mer to perform medical image analysis, and some quite complex work may be performed. However, the lack of programming does not mean that good practice associated with scientific rigor may be ignored. It is still necessary to justify the choice of operations and parameters. Information about how an operation has been implemented will be in the user manual for the software, or alternatively there may be online sources of information. A well-written manual will refer the reader to original publications on the technique. If no information is available, then it is advisable not to use the operation in a seri-ous study. It is sensible to check that an operation performs as expected on test data with known characteristics. Be aware also that even reputable com-mercially available software may not be licensed for clinical use, and should be used only for research purposes. Information about any such restrictions will be clearly stated.

### 10.5.2   Images on the Web

A beginner in image processing will probably download images from the Internet, perhaps from one of the many radiology teaching resources. The images will have been optimized for Web use, and are likely to be reduced in size and dynamic range compared with the original. The images will be compressed. Such images are good for practice, but should never be used for developing protocols, as they will have little in common with images that will be acquired for a study.

It is advisable to think carefully before placing images on the Web, keep-ing in mind privacy issues and the scope of any informed consent that was granted for the image. Illustrative Web images can be in the gif, jpeg or png formats, and are best optimized to reduce their size and thus the download time. It can be appropriate to watermark the image very obviously if repro-duction is to be discouraged. Copyright is often ignored online.

*Self-assessment question 10.01*

Joe has often seen it stated in articles that "images from n consecutive patients attending the radiology department were included in the study," so he decided to wait until all the images had been routinely acquired before considering the analytical protocol for his new study. What are the pitfalls associated with Joe's approach?

## 10.6 Evaluation of the Performance of Image Processing Methods

Three topics covered in earlier chapters—segmentation, registration and visualization—are image processing methods that require evaluation to determine that the result obtained is close to the truth. Evaluation may be either qualitative or quantitative. Qualitative techniques are most relevant to beginners in image processing, while full-blown quantitative evaluation is an essential part of professional software development.

### 10.6.1 Qualitative Evaluation

Qualitative methods of evaluating the results of image processing are analogous to the simple commonsense checks one makes when doing arithmetic.

#### 10.6.1.1 Qualitative Evaluation of Segmentation

A visual check of segmentation may be performed by comparing the result (test segmentation) with a reference standard. Binary images are used, with the test segmentation subtracted from the reference. The data type of the result of subtraction should be a signed one—in ImageJ, this is done by selecting the option for a 32-bit result. The image will then be gray for pixels that are the same in both images, negative (usually black) where present in the test but not in the reference, and positive (usually white) where present in the reference but not the test. An example is shown in Figure 10.3.

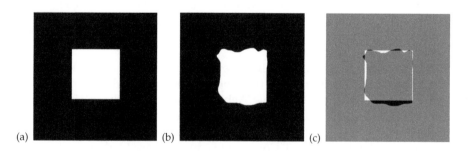

(a)           (b)           (c)

**FIGURE 10.3**
Example of a visual check of segmentation results using binary images with values 0 and 255. (a) Reference segmentation. (b) Test segmentation. (c) Result of subtraction of the test segmentation from the reference. In the 32-bit result, black represents –255, gray zero, and white +255.

### 10.6.1.2 Qualitative Evaluation of Registration

Similarly, a quick way to illustrate the presence of misregistration between a pair of images from the same modality is to subtract one from the other. If images were perfectly aligned, then the difference between them would be zero. Areas that do not match will appear brighter or darker than aligned regions. When a pair of images has been registered, it is expected that there will be some non-zero results, but there should be fewer after registration than there were before. A range of results is illustrated in Figure 10.4. The result of subtracting two identical images (Figures 10.4a and b) is seen in Figure 10.4c. The next two images differ by a small rotation (Figures 10.4d and e), and

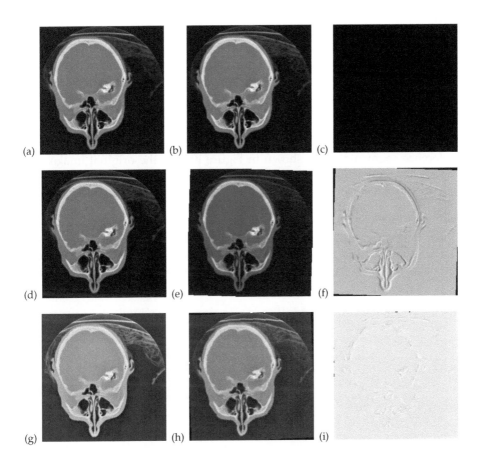

**FIGURE 10.4**
Example of a visual check of registration results for images from the same modality. (a)–(b) Identical base and match images, (c) result of subtraction is zero. (d)–(e) Base and match images are misaligned by a small rotation, (f) result of subtraction shows typical non-zero misregistration differences. (g)–(h) Base and roughly registered match images, (i) result of subtraction has fewer non-zero values than before registration. (Image data from the Chapel Hill Volume Rendering Test Data Set, Volume I. [1] With permission.)

typical misregistration effects are seen in the difference image (Figure 10.4f). Rough image registration has been performed in an attempt to realign the images (Figures 10.4g and h). In the difference image (Figure 10.4i), although there are fewer non-zero features, it can be seen that the registration has not been entirely successful.

### 10.6.1.3   Qualitative Evaluation of Visualization

The results of visualization have a strong reliance on the validity of the definition of the visualized object. In the case of scene-based methods, object definition depends on the choice of threshold values for the rendering. In object-based methods, there is a dependence on the accuracy of the segmentation that was used to generate the object. Qualitative methods for assessing segmentation results are therefore also applicable to visualization. Visual assessment of visualization results is highly subjective, and is further complicated by the lack of knowledge about the true anatomical appearance.

*Self-assessment question 10.02*

The result of subtracting a binary test segmentation from the corresponding reference segmentation is shown in Figure 10.5. In the original images, the object had gray value 255 and the background 0. Did the test segmentation give a segmented region systematically inside or outside the reference result?

*Self-assessment question 10.03*

Two images, resulting from the subtraction of two different pairs of images that have undergone image registration, are shown in Figure 10.6. Which image is associated with subtraction of the pair of images that are better registered?

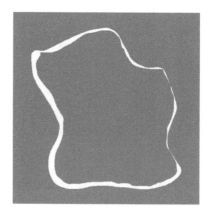

**FIGURE 10.5**
Image for self-assessment question 10.02. White indicates gray value +255, gray 0, and black −255.

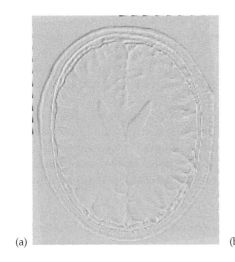 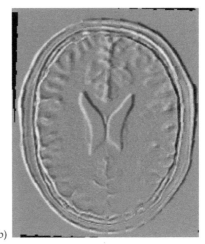

(a)       (b)

**FIGURE 10.6**
(a–b) Images for self-assessment question 10.02. (Image data from the BrainWeb Simulated Brain Database [2,3]. With permission.)

## 10.6.2 Quantitative Evaluation

The quantitative evaluation of an image processing method must take place before the method is applied to data in a clinical study, or is included in a commercial product. A full consideration of evaluation is an advanced topic beyond the scope of this introductory book, but because an understanding of the principles and terminology is valuable for anyone undertaking image processing, the topic is covered briefly here.

Quantitative evaluation concerns the *accuracy*, *precision*, and *efficiency* of the method. Accuracy is the term that defines how close the result is to the known correct answer, or ground truth. In many clinical examples, the ground truth is not known for certain. In those cases a "gold standard" is used instead, which is a value determined using what is believed to be the best method currently available for making a measurement.

Precision is also known as reproducibility or variability. Precision is a measure of the closeness of the answers from repeated measurements. A method can give a result far from the ground truth and still be precise if all the measurements give close to the same answer.

The assessment of reproducibility should cover the whole imaging process, including both the acquisition of the image and the analysis it undergoes. This is because there are a number of uncertainties in medical image processing that are unrelated to image analysis. There will be biological variability within a patient group, and different variability in normal and diseased populations. Images can vary from acquisition to acquisition, sometimes because of subject motion, but also because limited spatial resolution can result in variable partial volume effects associated with small positional differences. Overall, then, although automated analysis methods may give the same answer twice on the same image, to determine their true reproducibility the image analysis should be repeated on a different image from the same individual.

The third aspect of evaluation is efficiency, which concerns the time taken to undertake the processing, including both computer time and the time of the operator.

Evaluation is often performed using several iterations, in which the image analysis technique is refined to provide the best accuracy and precision on a set of data similar to that in the full study. Lessons from each iteration are incorporated into the revised software. This process of evaluation needs to be carefully planned to avoid an analysis method being finely tuned to work on a particular subset of data.

The issues associated with quantitative evaluation are slightly different for segmentation, registration and visualization.

### 10.6.2.1   *Quantitative Evaluation of Segmentation*

Image segmentation methods are very specific. Results of evaluation will be valid only for a particular clinical application, defined both by the type of task and the region of body, and the imaging modality with its specific acquisition parameters. Even if an algorithm is used that has previously undergone an evaluation, unless the published results are for exactly the same combination of clinical application and imaging protocol, it will be necessary to perform an evaluation.

Quantification needs to go beyond a count of the number of segmented pixels, as the pixels may have been identified in quite the wrong place in the image. There are different approaches dependent on the practical application. In some cases, clinically relevant landmarks are important, and the assessment will focus on determining how well the segmentation identified these. Where the border of a segmented region is important, evaluation is based on comparison of a test outline with a reference outline. The distance from each point on one curve to the closest point on the other is found. The mean of these closest-point distances may be used as an indication of agreement, but the mean is not a reliable way of picking up errors associated with only a few points in the outline (Figure 10.7). The Hausdorff distance between two curves, which is the maximum closest-point distance, does not have this drawback, and so is a popular comparator.

For comparisons of segmented regions, as opposed to boundaries, an approach adapted from assessment of diagnostic performance can be used. This approach is described more fully because it is feasible to implement the analysis in ImageJ. Diagnostic performance is often expressed in terms of the sensitivity (true positive fraction) and specificity (true negative fraction) of the test. [4] A true positive result is when the diagnostic test correctly identifies subjects that gave positive results using the reference test. A false positive is when the diagnostic test classes a subject as positive when a positive

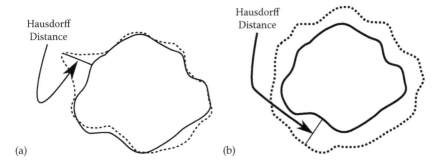

**FIGURE 10.7**
The Hausdorff distance (maximum closest-point distance) is a better measure than the mean closest-point distance for indicating errors associated with only a few points in the outline. The mean closest-point distance is smaller in (a) than in (b), but the Hausdorff distance is about the same size in both cases.

result was not obtained using the reference test. For evaluating segmentation algorithms, the same terminology may be applied in a slightly different way. Instead of referring to diagnosis of disease, a true positive result is simply when a pixel is found by the segmentation method to be part of the object when this was also the case for the reference segmentation. The principle is illustrated in Figure 10.8. Note that the segmented image may contain several separate objects, and an extension to the method is required if it is important for segmented pixels to be associated with a particular object.

The 2 × 2 table is the conventional way to present the results of testing a diagnostic test against a reference standard. The same kind of table may be used when evaluating segmentation methods—a pixel is described as positive if it is within the segmented region, and negative if outside. It is shown

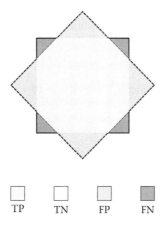

**FIGURE 10.8**
Diagram showing the definitions of true positive (TP), true negative (TN), false positive (FP), and false negative (FN) in the evaluation of segmentation results. The segmentation (larger, rotated square) is compared with a reference segmentation (smaller square) using logical operators.

**TABLE 10.1**

2 × 2 Table for Evaluation of Segmentation Results

|  | Reference Positive | Reference Negative |
|---|---|---|
| Positive in test segmentation | True positive (TP) | False positive (FP) |
| Negative in test segmentation | False negative (FN) | True negative (TN) |
| Total | P | N |

Sensitivity is defined as the True Positive Fraction = TP/P

Specificity is defined as the True Negative Fraction = TN/N

in Table 10.1 how the different combinations allow each pixel in a segmented image to be defined as true or false positive, or true or false negative. The definitions of sensitivity and specificity in terms of P, N, TP, and TN are also given in the table.

Sensitivity and specificity can be determined using values from the binary images as follows:

- The number of positive pixels in the reference (P) and the number of negative pixels in the reference (N) may be counted without further processing.
- AND, when applied to the reference and test images, will give an image showing the true positive pixels (TP).
- AND applied to the two images after grayscale inversion will give an image showing the true negative pixels (TN).

The performance of a test in terms of sensitivity and specificity is often illustrated using the *Receiver Operating Characteristic* (ROC) curve. [5] ROC curves are also useful for comparing the performance of different tests. A ROC curve (Figure 10.9) plots sensitivity against (1–specificity). The results for a perfect test, with 100% sensitivity and specificity, would be plotted at the top left corner. Curves more distant from the top left corner of the graph represent worse performance. For the two example curves shown in Figure 10.9, the test plotted with the solid line has the better performance.

### 10.6.2.2 Quantitative Evaluation of Registration

As for image segmentation, any evaluation of image registration will be specific to the particular clinical application and the imaging acquisitions involved. Quantitative comparison with a reference registration may be made either by comparing the registration transformations, or by determining the distance between the locations of important anatomical features calculated for each transformation. The main difficulty in making such a comparison is that the reference registration may not itself be very good. It is very challenging to find an appropriate reference standard, especially for nonrigid registration. It is tempting to generate misregistered images at

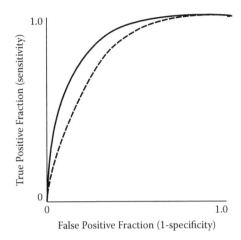

**FIGURE 10.9**
Typical form of receiver operating characteristic (ROC) curves. The test represented by the results shown with the solid line has the better performance.

known transformations from their reference positions by using the registration software in reverse. This is a pitfall to avoid, because the misregistered images produced in this way cannot be considered to be completely independent of the software under test. They have been generated using a particular method, and so it would be expected that they will be readily realigned using that particular method. As a result, the method is not properly challenged.

A more robust approach was taken in a well-known study. [6] The reference registration transformation was defined by using fiducials in the acquired images. The image sets were processed to remove the fiducials, and were then used by many research groups who applied their algorithms to the data and compared their results with the reference results. An article on this study is the basis for one of the case studies in Chapter 11. An approach that does not rely on fiducials or landmarks is the consistency approach. The method requires three independently acquired images A, B, and C. For a perfect registration method, the result of registering A directly to C will be the same as registering A to B and then to C. In practice, there will be a difference in the results, and, by making assumptions regarding the statistics of the errors, they can be quantified.

### 10.6.2.3   Quantitative Evaluation of Visualization

Evaluation of images produced by visualization methods is made difficult by the large number of parameters related to both acquisition and processing that can affect the results. Evaluation may be done in terms of technical image quality or, if the visualizations are used for diagnostic purposes, in terms of the accuracy of the diagnosis made using the images. The former is the appropriate approach when a technique is under development, and the effect of different values for parameters is of interest. The reference data used for comparison could be an anatomical specimen, a specially designed test object or phantom, or computationally simulated data. The most obvious drawback of physical phantoms is their lack of anatomical shape; phantoms are often designed geometrically to ensure that particular sizes and shapes are included. If segmentation is performed as part of the visualization technique, then in the absence of agreed methods for the evaluation of the results of visualization, evaluation of the segmentation is a robust first step.

For further reading on evaluation see the overview by Bowyer, [7] and more specific articles by Udupa et al. on segmentation, [8] Fitzpatrick on registration, [9] and Pommert and Höhne on visualization. [10]

### 10.6.3   Health Technology Assessment

The evaluations discussed in this chapter were focused on testing the performance of the image analysis algorithms themselves. This type of evaluation falls in the first layer of a hierarchy of tests associated with assessment of the efficacy of diagnostic methods: [11]

- Technical Performance
    - Are images of sufficient technical quality; are they anatomically accurate and are analysis protocols accurate, precise and efficient?
- Diagnostic Performance
    - Is the diagnosis made using the images correct? This is usually expressed in terms of sensitivity and specificity.
- Diagnostic Impact
    - Do the results change diagnostic confidence? Are other tests displaced by the new one?
- Therapeutic Impact
    - Do the results change how the patient is treated?
- Impact on Health
    - Is there a measurable change to the patient's health as a result of having the diagnostic test?
- Societal Impact
    - What are the broader economic implications of using the diagnostic test?

Results from the evaluation of image analysis techniques need to be considered in the context of this wider view, as an improved image analysis algorithm does not necessarily translate into improved performance at the higher levels. Furthermore, image analysis is increasingly used in areas other than diagnosis, such as monitoring and for the guidance of therapeutic interventions.

## 10.8 Chapter Summary

This chapter focused on good practice in image processing. The points raised in earlier chapters were grouped together and discussed in general terms in relation to scientific rigor, ethical practice, human factors and information technology. Qualitative and quantitative evaluation methods for image processing algorithms were outlined, with the emphasis on segmentation, registration and visualization.

## 10.9 Feedback on Self-Assessment Questions

### Self-assessment question 10.01

If all the images are routinely acquired before considering the analytical protocol, the pitfalls include the following. There is a likelihood that the images will not all be acquired using the same acquisition protocol, and variations in slice thickness and other parameters can lead to difficulties in performing comparable analyses. Control acquisitions should be made at the same time as the subject data are acquired, to offset the effect of large changes caused, for example, by the installation of new equipment. If images are acquired using standard clinical protocols, they may be difficult to analyze. It is always sensible to plan ahead and ensure that an acquisition protocol suitable for later analysis is in place from the outset, including consideration of both technical aspects and patient positioning. Without specific instructions to the contrary, the acquired images may have been compressed using a lossy technique.

### Self-assessment question 10.02

The result has no pixels with value −255, meaning that the test segmentation was inside the boundary of the reference segmentation. Note that automatic windowing could mean that a subtraction result like this, with only two gray levels, would be displayed using black and white to represent 0 and 255.

### Self-assessment question 10.03

The better registered pair is pair (a). There are fewer misregistration features visible in the image, and those present are less strongly black or white, meaning that the differences between the images are smaller.

## References

1. The Chapel Hill Volume Rendering Test Data Set, Volume I, Department of Computer Science, University of North Carolina. Available at http:// education.siggraph.org/resources/cgsource/instructional-materials/volume-visualization-data-sets.
2. The BrainWeb Simulated Brain Database. www.bic.mni.mcgill.ca/brainweb/
3. Collins, D.L. et al. Design and construction of a realistic digital brain phantom, *IEEE Trans. Med. Imag.*, 17, 463, 1998.
4. Greenhalgh, T. How to read a paper: Papers that report diagnostic or screening tests, *BMJ*, 315, 540, 1997.
5. Obuchowski, N. A. Receiver operating characteristic curves and their use in radiology, *Radiology*, 229, 3, 2003.
6. West. J. et al. Comparison and evaluation of retrospective intermodality image registration techniques, *J. Comp. Assist. Tomog.*, 21, 554, 1997.
7. Bowyer, K.W. Validation of medical image analysis techniques, in *Handbook of Medical Imaging Volume 2. Medical Image Processing and Analysis*, Sonka, M. and Fitzpatrick, J.M., Eds., SPIE Press, Bellingham, 2000, 567.
8. Udupa, J.K. et al. A framework for evaluating image segmentation algorithms, *Comput. Med. Imag. Grap.*, 30, 75, 2006.
9. Fitzpatrick, J.M. Detecting Failure, Assessing Success, in *Medical Image Registration*, Hajnal, J.V., Hill, D.L.G., and Hawkes, D.J., Eds., CRC Press, Boca Raton, 2002, 117.
10. Pommert, A. and Höhne, K.H. Evaluation of image quality in medical volume visualization: The state of the art, in *Medical Image Computing and Computer Assisted Intervention, Proc MICCAI 2002, Part II*, Lecture Notes in Computer Science 2489, Takeyoshi, D., and Kikinis, R., Eds., Springer-Verlag, Berlin, 2002, 598.
11. Mackenzie, R. and Dixon, A.K. Measuring the effects of imaging—an evaluative framework, *Clin. Radiol.*, 50, 513, 1995.

# 11

## Case Studies

### 11.1 Introduction

This chapter includes five case studies, each based on a published journal article. The case studies include questions on the article itself and activities using ImageJ, in which image processing related to the subject of the article is performed. Feedback is provided at the end of each case study. There are two ways in which this chapter is one of consolidation. Firstly, it helps the learner to bring together, and practice, critical skills in extracting relevant information from published material and newly developed practical image processing skills. Secondly, it shows how the topics covered in separate chapters are used together when tackling real-world projects.

#### 11.1.1 Learning Objectives

When you have completed this chapter, you will

- Have further developed your skills in critically reading research articles
- Have enhanced your ability to communicate the information presented in an article without quoting it directly
- Have gained further experience in using the ImageJ image processing program to enhance, segment, compress, align and visualize medical images
- Have an understanding of factors to consider when planning a research study involving image processing

### 11.2 Image Enhancement

The article [1] required for this case study is also available online without subscription. [2] The images required for this case study are on the CD.

## 11.2.1   Questions on the Image Enhancement Article

Read the article [1,2] and write down answers to the following questions, consulting the article when necessary.

### Question 11.2.1.1

What advantages are given in the article regarding the use of a digital frame of reference (DFOR)?

### Question 11.2.1.2

Choose two examples given in the article of image processing techniques that may result in image artifacts (appearances that could mislead). Suggest two more that are not given in the article.

### Question 11.2.1.3

The images in Figure 11.1b–h show the results of applying different image processing operations to the reference image shown in Figure 11.1a. Seven image processing operations are listed below: Match one operation to each image, explaining your reasoning. Once you have thought about the question, use ImageJ to check the answers by applying each operation to referenceimage.bmp (which you will find on the CD). Take care to load referenceimage.bmp before performing each operation, so that you are sure that only the one operation has been performed.

- Window width reduced from 255 to 100, window level 128 (as before)
- Inversion of look-up table
- Mean filtering using a kernel of radius 5 pixels
- Median filter with kernel of radius 5 pixels
- Edge detection, using ImageJ's Find Edges option, which is Sobel filter based
- Division by 2
- Thresholding to set pixels with values in the range 192–255 to 255, and others to zero

## 11.2.2   Activities: Image Enhancement and the Digital Frame of Reference Using ImageJ

In a conventional hip replacement operation, the head of the femur is removed and replaced with a prosthesis. The hip replacement consists of a metal ball that is attached to a shaft, which is fixed inside the femur. Over time, wear to the prosthesis means that it may become loose, and many need replacement after 10 or more years. X-ray imaging is the modality used to

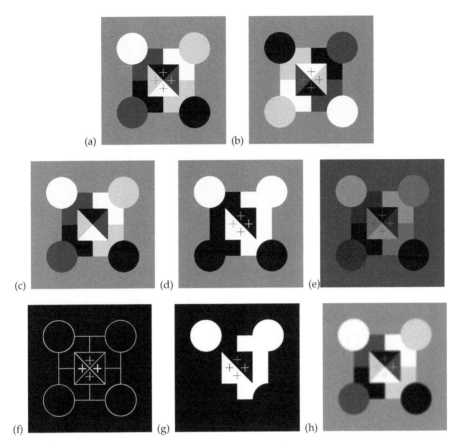

**FIGURE 11.1**
Images for question 11.2.1.3. (a) Reference image, (b)–(h) results of applying different image processing operations to the reference image.

check the integrity of the hip replacement. The main area of interest is the border between the prosthesis and bone, where signs of bone deterioration or bone loss indicate loosening of the prosthesis in the femur. It is typical for the sharpness of the image to be enhanced before the image is assessed, and in this activity the effects of the unsharp mask filter (Chapter 3) are demonstrated.

Load hip_replace.tif from the CD into ImageJ (Figure 11.2a).

Process | Filters | Unsharp Mask. In the dialog box, set Gaussian Radius to 2, Mask Weight to 0.8, and click OK.

Load another copy of hip_replace.tif and make a visual comparison of the two images, especially around the tip and sides of the prosthesis, noting your comments.

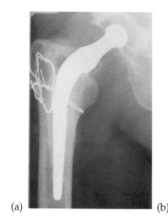
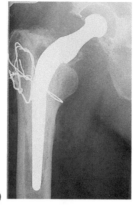
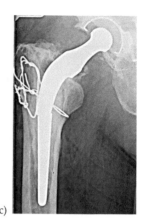

(a)                              (b)                              (c)

**FIGURE 11.2**
(a) Original image. (b) Unsharp masking with Gaussian Radius 2 and Mask Weight 0.8. (c) Unsharp masking with Gaussian Radius 10 and Mask Weight 0.8.

Select the unprocessed copy of hip_replace.tif and perform unsharp masking as before, but this time set the Gaussian Radius to 10 and Mask Weight to 0.8.

Load another unprocessed copy of hip_replace.tif and comment on the different appearances of the three images.

In the first unsharp masked result (Figure 11.2b), edges appear stronger than in the original image but noisy patterns, which may be artifactual, have appeared in the bone and within the prosthesis. However, these changes do not adversely affect the appearance of the boundary of the stem of the prosthesis, which is the area that will be assessed for loosening. In the image that has had more sharpening applied (Figure 11.2c), there is a black halo around bright structures, including the prosthesis, which is especially noticeable at the tip. The lucent rim around the prosthesis simulates bone degradation or bone loss and so implies that the prosthesis is loose. This effect could be a problem if the radiologist viewing the image was unaware of its processing history. It was suggested in the article that this is just the sort of situation in which a digital frame of reference would be of value.

Use ImageJ to perform the same processing on referenceimage.bmp as you did on the prosthesis image. Note the more pronounced haloing effect at every edge in the image filtered using the larger value for the Gaussian radius. If, as is suggested in the article, the observer were presented with a reference image that had undergone the same processing as the diagnostic image, suspicions would be aroused and the original image would be examined to determine if the apparent bone loss were real or an artifact of image processing.

### Psychophysical Studies

The article concludes by suggesting that the digital frame of reference should be tested in a clinical application by means of a psychophysical experiment. Check, using a dictionary or the Internet, that you understand what a psychophysical experiment is. Consider the factors that it will be important to take into consideration when designing the suggested psychophysical study in order to avoid bias in the results.

## 11.2.3   Feedback on the Image Enhancement Case Study

### *Question 11.2.1.1*

As more diagnostic images are acquired digitally, it is more likely that some form of image processing will be applied to the image before it is viewed by the diagnostician. If the observer is unaware of the processing history of the image, then there is potential for misdiagnosis arising from artifactual features that have been introduced by the processing. The digital frame of reference (DFOR) is a way of recording what has happened to an image and presenting that information to the observer in a visual manner.

### *Question 11.2.1.2*

Three examples are given in the article: unsharp masking leading to a halo effect at edges, contrast or look-up table inversion leading to the exchange of black and white areas, and the sharpening filter.

There are many further possibilities as most operations, especially if the values for the parameters are badly chosen, can change an image considerably:

- In extreme cases of contrast windowing, areas of the image saturate (i.e., are displayed in black or white). There is the possibility that using windowing to enhance one feature means that another important feature is lost.
- If blurring, smoothing and averaging filters are used, small-scale features may disappear. It would not be valid to assess texture on images that had been blurred.
- Jpeg compression introduces artifactual dots that could be interpreted as features, and may be enhanced by further processing.
- If the purpose of imaging were to detect small-scale features much brighter or darker than the surroundings, it would not be useful to apply a median filter or to smooth the image.

**Question 11.2.1.3**

- Window width reduced from 255 to 100, window level 128 (as before) is seen in Figure 11.1d. Only pixels with values between 78 and 178 retain their original values; those outside the range are set to zero or 255.

- Inversion of look-up table is seen in Figure 11.1b; black areas are now white and white black.

- Mean filtering using a kernel of radius 5 pixels is seen in Figure 11.1h, the only image which looks blurry.

- Median filter with kernel of radius 5 pixels is seen in Figure 11.1c. Pixels in boundary regions and in areas smaller than the kernel size have been set to have a value the same as the surroundings.

- Edge detection, using ImageJ's Find Edges option, which is Sobel filter based. This is seen in Figure 11.1f, which has the sharp changes in intensity highlighted.

- Division by 2 is seen in Figure 11.1e, which is a dimmer version of the original, with no saturation where pixels are set to 0 or 255.

- Thresholding to set pixels with values in the range 192–255 to 255 and others to zero. This is seen in the binary image Figure 11.1g.

---

## 11.3   Segmentation

The article [3] required for this case study is also available online without subscription [4]. The images required for this case study are on the CD.

### 11.3.1   Questions on the Segmentation Article

Read the article [3,4] to get an overview of what the article is about, and make a mental note of where in the article different information is provided. Consult the article when necessary and write down answers to the following questions.

**Question 11.3.1.1**

What are the four reasons given in the article to explain why results for the volume of the hippocampus and amygdala from different laboratories may not be directly comparable? Take care not to quote text directly from the article in your answer.

**Question 11.3.1.2**

On page 434 of the article, in the section on MR image analysis, three steps are listed in the text to outline the processing that was performed on the image volumes before segmentation took place. These are all issues that have been covered earlier in this book: (a) correction for magnetic field nonuniformities, (b) linear stereotaxic transformation into coordinates based on the Talairach atlas and (c) resampling onto a 1 mm voxel grid using a linear interpolation kernel. Explain the need for each of the steps (a) and (b). As an example, the need for step (c) may be explained as follows. When a transformation is applied to an image volume, the relative locations of the voxels change and they may no longer lie neatly and equally spaced on a rectangular grid. Interpolation is necessary to ensure that the new image is made up of voxels that are in the regular arrangement expected for further processing.

### Question 11.3.1.3

Explain how interobserver and intraobserver agreement was assessed in the study.

### Question 11.3.1.4

On page 438 of the article, in the second paragraph of the Results section, it is stated that "interaction between gender and hemisphere for the HC was not significant." In the fourth paragraph it says "... also showed a significant gender difference." So, there is a significant gender difference for one set of results (paragraph four) but not the other (paragraph two). Results for two different analytical approaches are presented in the two paragraphs; what is the difference between these two approaches? Based on this knowledge, does it seem reasonable that the significant gender differences arise for the second analysis?

### 11.3.2 Activity: Manual Segmentation Using ImageJ

The case study article concentrated on manual segmentation of two brain structures, the hippocampus and the amygdala. Fully automatic segmentation would be very challenging because of the complexities of the anatomy, small size and low image contrast. However, semiautomatic segmentation is attractive for this application as it would combine the repeatability of automated analysis with the necessary operator knowledge of the shape, location and boundary characteristics of the structures. In the following activities, manual and semiautomated segmentation methods are used, but because the hippocampus and amygdala are so complex to define, to make things simpler a larger structure of the brain is the subject of the segmentation.

Start ImageJ. Ensure that ImageJ is set up to display correctly using Edit | Options | Colors. Set Foreground to white and Background to black, as in Figure 11.3.

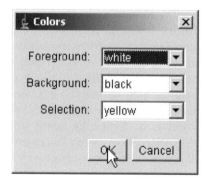

**FIGURE 11.3**
The ImageJ Colors dialog.

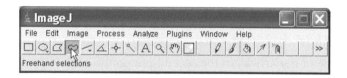

**FIGURE 11.4**
The ImageJ freehand selections icon in the ImageJ Tools menu.

Select File | Open and open the image PDw.tif* from the CD.

Click on the freehand selections button (Figure 11.4).

Draw carefully around the butterfly-shaped brain ventricles (Figure 11.5a). Draw this outline as one shape. If you try to do two separate ones, the first will disappear.

Select Edit | Clear outside to give an image like the one in Figure 11.5b.

Select Edit, Selection, Select None to remove the drawn line.

To convert the result into a binary image, select Image | Adjust | Threshold.

Adjust the threshold range (Figure 11.6a) so that all the pixels in the ventricles are selected (Figure 11.6b).

---

* Image data from the BrainWeb Simulated Brain Database [5,6]. With permission.

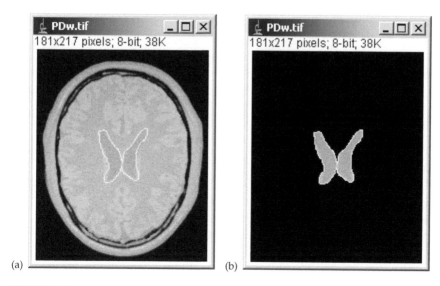

**FIGURE 11.5**
Segmentation of the ventricles. (a) Draw an outline. (b) After Clear outside operation. (Image data from the BrainWeb Simulated Brain Database [5,6]. With permission.)

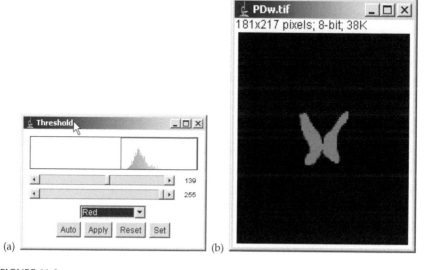

**FIGURE 11.6**
Thresholding the selection. (a) The ImageJ Threshold dialog. (b) The selected pixels, which appear gray in this figure, are shown in red in ImageJ. (Image data from the BrainWeb Simulated Brain Database [5,6]. With permission.)

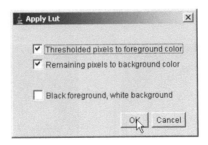

**FIGURE 11.7**
The ImageJ Apply LUT dialog.

Click on Apply, select the settings shown in Figure 11.7, and click on OK. Close the threshold box. Don't worry if the ventricles go red again.

Save the thresholded image in tif format under a new name, for example, your_initials_seg_PDw.tif

### 11.3.3  Activity: Comparison with a Binary Reference Standard Image Using ImageJ

In practical studies, it is necessary to compare results of segmentation with a reference segmentation. In a few cases, for example, with synthetic data and phantoms, this is the ground truth. More often, the reference will be the best available segmentation, known as the gold standard. In this activity, the reference segmentation is supplied as a binary image. Before continuing with the image processing, the range of possible outcomes from subtracting a pair of binary images needs to be considered. In the binary images, segmented pixels have value 255 and the background is 0. The various combinations that are possible in a pair of binary images of the same matrix size are:

- 255 in reference, 255 in the other image
- 255 in reference, 0 in the other image
- 0 in reference, 255 in the other image
- 0 in reference, 0 in the other image

If the other image were subtracted from the reference image when the pixel value in both images was 255, then the result would be 0. This result could be described as a True Positive, because the pixel has been defined as being part of the object in both the segmented image and in the reference. Before reading the next paragraph, write down the numerical results expected from subtraction of the other image from the reference image for the other three combinations of pixel values. For each, assign the correct description from False Positive, True Negative, and False Negative.

The result of subtraction will be an image with up to three possible pixel values: −255 for false positives, +255 for false negatives, and 0 for both true positives and true negatives. Conceivably, for a very good subtraction, the result might include only two different values. For example, a segmentation with no false positives would have values 0 and 255 only.

Close any open ImageJ windows, then load the recently created file (your_initials_seg_PDw.tif) and the reference segmentation (justventriclesthresholded.tif).

Select Process | Image Calculator and set up the operation as shown in Figure 11.8. Click on OK. The result should be similar to the image shown in Figure 11.9. If the result does not have a gray background with black and

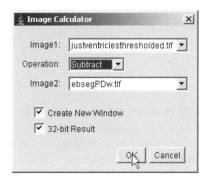

**FIGURE 11.8**
The ImageJ Image Calculator dialog.

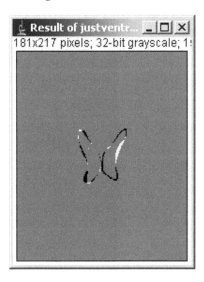

**FIGURE 11.9**
Result of subtraction of test segmentation from the reference segmentation. (Image data from the BrainWeb Simulated Brain Database [5,6]. With permission.)

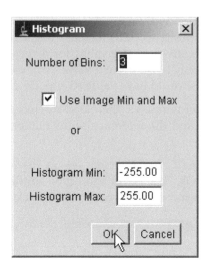

**FIGURE 11.10**
The ImageJ Histogram dialog.

white pixels, check first that the box was ticked for a 32-bit result, which is a way of ensuring that both positive and negative values are seen in a result (Chapter 5). Alternatively, the segmentation may have been very good indeed, so that there are no values of either −255 or 255. This will lead to a black and white image with no grays. Move the cursor over the pixels to see their values—a value of zero should be present in the background area.

The next step is to count up the number of black (pixel value −255) and white (pixel value +255) pixels. This is achieved using the histogram option. Click on the result image to ensure that it is the one analyzed, then select Analyze | Histogram. Set up the histogram as in Figure 11.10 to have up to three bins, and click OK. The result should look like Figure 11.11a. Click on List, and a list of how many pixels are in each range is returned (Figure 11.11b). Note the number of false positive and false negative pixels, preferably laid out in a table as indicated in Chapter 10 (see Table 10.1).

### Histogram Bins

Are you concerned about whether or not the "bin start" values in the histogram make sense? The range of pixel values is −255 to + 255, that is, there are 511 values. If three sections (bins) are required in the histogram, then each should cover 511/3 = 170 values. The bin ranges will therefore be −255 to −85, −85 to +85 and +85 to +255, so the bins are just as expected.

### 11.3.4   Activity: Reproducibility of Manual Segmentation

To investigate intraobserver reproducibility repeat the segmentation of PDw.tif and the comparison with the binary reference image. Note the

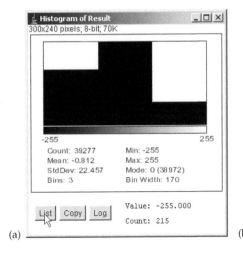

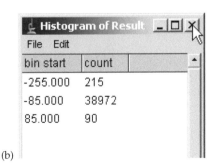

(a)  (b)

**FIGURE 11.11**

(a) Result of Analyze | Histogram in ImageJ. (b) Numerical histogram results obtained by using the List option.

number of false positive and false negative pixels. If time permits, the segmentation could be performed using the images acquired using different acquisition sequences (T1w.tif and T2w.tif are on the CD) to determine if there is a difference, which exceeds the difference arising from intraobserver reproducibility, as a result of using a different image.

### 11.3.5 Activity: Semiautomated Segmentation Using ImageJ

The semiautomated technique used in this activity will be one of the simplest semiautomated segmentation techniques: gray-level thresholding.

Using ImageJ, close open windows, then load T2w.tif* from the CD.

Select Image | Adjust | Threshold (Figure 11.12). Adjust the upper and lower values until all the pixels in the ventricles are red. Other areas of the image will be selected, too (Figure 11.13a). Make a note of the range of values used. Click on Apply (Figure 11.13b).

Use the Freehand Selections button (Figure 11.13c), and the Clear outside function to retain only the ventricles (Figure 11.13d). Save the thresholded image in tif format under a new name, for example, your_initials_segthresh_T2w.tif.

Repeat the image subtraction procedure to compare the result with the binary reference image (justventriclesthresholded.tif), and note the number of false positive and false negative pixels.

---

* Image data from the BrainWeb Simulated Brain Database [5,6]. With permission.

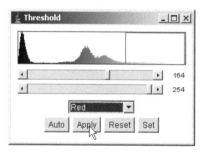

**FIGURE 11.12**
The ImageJ Threshold dialog.

### 11.3.6    Questions about the ImageJ Activities

*Question 11.3.6.1*

Were there any systematic differences between the semiautomatic and manual segmentations?

*Question 11.3.6.2*

Do you think that choosing the threshold by eye, as was done in the activity, is a good way to choose the threshold if reproducible results are required? Give a reason for your answer.

*Question 11.3.6.3*

In question 11.3.1.1, you were asked to list the reasons given in the article for variations in measurements between laboratories. Look back to the answer you gave there, and explain here which aspects are relevant to the practical activities just performed.

### 11.3.7    Feedback on the Segmentation Case Study

*Question 11.3.7.1*

Four reasons are given in the article to explain why results for the volume of the hippocampus and amygdala from different laboratories may not be directly comparable. Firstly, different image acquisition sequences may have been used. Even if the most common method, the T1-weighted sequence, is used, the protocols may still differ because of the range of values used for parameters such as TE, TR, slice thickness and pixel size. These factors will affect measured volume. Similarly, if three-dimensional data sets with equal resolution in all directions are available, then they are expected to give different results from those using 2-D slices. Not all groups use manual segmentation. Results from automated techniques may have a systematic difference from the results acquired using manual methods. Finally, the definition of the anatomical structure may vary from laboratory to laboratory.

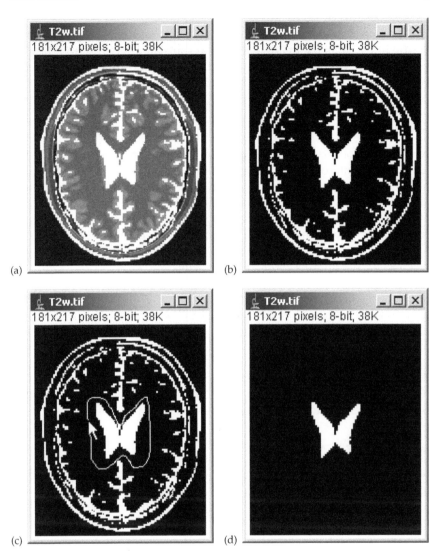

**FIGURE 11.13**

(a) Selected pixels (white in this image) are usually shown in red in ImageJ. (b) After apply-ing the chosen threshold. (c) Using freehand selections to select the ventricles. (d) After using Clear outside to retain only the ventricles. (Image data from the BrainWeb Simulated Brain Database [5,6]. With permission.)

## Question 11.3.7.2

(a) Correction for magnetic field nonuniformities is required because a lack of uniformity across the field of view can lead to gray-level variations in MR images. The appearance of a particular tissue depends on its location in the image. Where image gray levels are important (e.g., for segmentation), it is usual to correct for these inhomogeneities.

(b) The linear stereotaxic transformation into coordinates based on the Talairach atlas means that the images are all aligned with a brain atlas. The various structures will then be found in a standardized position and the volumes are corrected for head size.

### Question 11.3.7.3

Inter- and intraobserver agreement are called inter- and intrarater reliability in this article. Interobserver agreement was assessed using four observers, who each measured five of the subjects once. Intraclass correlations were calculated to give reliability coefficients, and the differences between the observers' measurements were expressed as percentages. Measurement of intraobserver agreement involved one of the observers analyzing the five subjects five times, at intervals of one week. The same analysis was performed as for the interobserver measurements.

### Question 11.3.7.4

The results in paragraph two are for the data that have been matched to the Talairach atlas, where a correction has been performed for head size. In paragraph four, however, "native space" is used. In other words, the mapping has not been performed and so any differences in head size will be retained. As male heads on average are larger than female, a gender difference for the brain structures would be expected in this case.

### Question 11.3.7.5

The response here is dependent on the individual's method of defining the boundary in the manual segmentation and the threshold for the semiautomatic segmentation. It is common, for example, for the outlines from thresholding to be wholly within the manually drawn boundaries.

### Question 11.3.7.6

No, choosing the threshold by eye is not a good way to choose the threshold if reproducible results are required. The approach adds subjectivity to what should be an objective method. In a research study, one would need to find a reproducible way of defining the range to use.

### Question 11.3.7.7

The issues mentioned in the article which were seen in the ImageJ activities were the dependence on the acquisition sequence, the segmentation technique and the individual's definition of the ventricular outline.

## 11.4 Image Compression

The article [7] required for this case study is also available online to subscribing institutions. The images required for this case study are on the CD.

### 11.4.1 Questions on the Image Compression Article

The authors of the article [7] have compared a number of compression schemes and general-purpose compression programs. They used example medical image data sets, because performance can be strongly related to the data characteristics. Read the article, paying particular attention to the Abstract and Introduction, as these will provide the information needed to answer the questions. Note that file sizes given in Table 1 and Table 9 in the article are stated in bytes (abbreviated to "b" in the article; here we use B to represent bytes). There are 8 bits in a byte, that is, 8 bits per pixel = 1 byte per pixel.

### Question 11.4.1.1

You are planning to send some medical images as an e-mail attachment. However, you know that your Internet Service Provider (ISP) does not allow attachments bigger than 1 MB in size, so you wish to compress the file first to a size that is within the limits. To perform compression, you have access to a copy of the general-purpose compression program PKZIP, which is one of those tested by Kivijärvi et al. in their article. Note that in their comparison tables, PKZIP is abbreviated to ZIP. From Table 9 in the article, what is the compression ratio for PKZIP on CT HEAD images?

### Question 11.4.1.2

Using the data in Table 9 on file sizes, would it be possible to e-mail the example set of 10 CT HEAD images as a single compressed e-mail attachment, using your ISP?

### Question 11.4.1.3

Ignoring the header size, and assuming that each slice occupies 524,288 bytes (called the real size in the article), what is the maximum number of CT HEAD slices that you could e-mail as a single attachment after compression with PKZIP?

### Question 11.4.1.4

Would it be possible to send the example NM BRAIN data in the article as a single compressed e-mail attachment, using your ISP?

### Question 11.4.1.5

What features of the two modalities would explain the difference in the amount of compression achieved for the CT HEAD and NM BRAIN data.

### Question 11.4.1.6

Using the data given in Tables 10 and 3, decide which general-purpose compression program, which runs under Windows 95, is the most efficient for compressing MR ABDOMEN images.

Lossy jpeg compression was discussed in Chapter 6, and it was seen in an activity that the lossy compression method changed pixel values compared with the original image. In the following activities the effects of lossy jpeg compression are considered further, and the lossless PNG image format is also studied.

### 11.4.2   Activity: File Size vs. Real Size

In this activity, the difference between the terms File size and Real size, as defined in Table 1 of the article [7], is addressed.

Use Windows Explorer®* and select the file EBLTestCard.raw on the CD. Open a menu of options using the right mouse button, and select the Properties option. Note the file size in bytes (not the file size on disk, which is somewhat dependent upon how the computer is set up). Repeat for EBLTestCard. tif. You should see that the file EBLTestCard.raw is 93,312 bytes in size and EBLTestCard.tif is 94,080 bytes.

Start ImageJ.

Select File | Import | Raw … and choose EBLTestCard.raw.

Complete the Import dialog. The image is an 8-bit image, 288×324 pixels in size. The offset to the first image is 0, there is one image and the gap between images is 0. White is not zero, the byte order is not little-endian and all files in the folder should not be opened. Click on OK.

Select File | Open and select EBLTestCard.tif.

Both images (Figure 11.14) are labeled as being 288×324 pixels and 8-bit, which means that for both image files the Real Size is 288×324 = 93,312 bytes. This is the file size noted for EBLTestCard.raw, so we can conclude that the .raw

---

* Registered trademark of Microsoft Corporation, Redmond, Washington, USA.

**FIGURE 11.14**
The EBLTestCard image, which is an 8-bit image of 288×324 pixels.

image file contains only image data and no header. On the other hand, the tif format file includes 768 bytes of header data in addition to the image data.

Notice that the Real size displayed above the images in ImageJ is 91K. The uppercase letter K represents 1024 bytes, so the size in kilobytes (KB) is obtained by dividing the number of bytes by 1024.

Close both open images.

### 11.4.3 Activity: Lossy jpeg Artifacts

This activity builds on the activity at the end of Chapter 6. A specially designed test image is used to emphasize all four jpeg effects:

- Blocking, because of the use of 8×8 pixel areas
- Blurring, because high-frequency components have been discarded
- Ringing at edges, because high-frequency components have been removed using a sharp cut-off frequency
- Changes to pixel values

Start ImageJ and open the image file EBLTestCard.tif.

Select Edit | Options | Input/Output.

Change the value in the box labeled "JPEG Quality (0–100)" to 12 and click OK.

Select File | Save As | Jpeg …, and save the image as EBLTestCard _12.jpg. Close the image by clicking on the cross at the top right.

Open the image file EBLTestCard.tif again.

Select Edit | Options | Input/Output. Change the value in the box labeled "JPEG Quality (0–100)" to 25 and click OK.

Select File | Save As | Jpeg ..., and save the image as EBLTestCard _25.jpg. Close the image.

Repeat to save two further jpeg versions of the tif image with image qualities 50% and 100%.

Open all four jpeg files. Move the images on the desktop until you can compare them visually side by side. If the images are not displayed full size, click on each image in turn and use Image | Zoom | View 100%. This means that any additional artifacts arising from scaling the images for display will not be seen.

The increasing degradation in image quality is visible (Figure 11.15).

Answer the following two questions, then close any open images.

### Question 11.4.3.1

Identify features in the images that show blocking, blurring, ringing and changed pixel values.

### Question 11.4.3.2

Calculate the compression ratio for each of the four jpeg images using Equation 1 given in the article. Note that the original image size in this expression is the Real size before compression, and the compressed image size is the File size after compression.

### 11.4.4 Activity: Effect of Repeated Jpeg Saves

Open the image file EBLTestCard_50.jpg.

Select Edit | Options | Input/Output.

Change the value in the box labeled "JPEG Quality (0–100)" to 50 and click OK.

Select File | Save As | Jpeg ..., and save the image as EBLTestCard _50_50.jpg. Close the image by clicking on the cross at the top right.

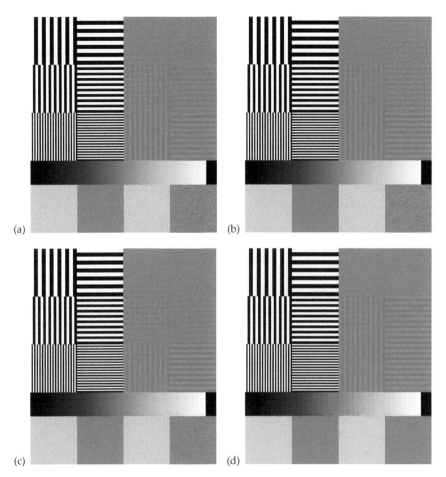

(a)      (b)

(c)      (d)

**FIGURE 11.15**
Effect of jpeg image quality setting. (a) 100%, (b) 50%, (c) 25%, (d) 12%.

## Question 11.4.4.1

Use Windows® Explorer to determine the size of the files EBLTestCard_50.jpg and EBLTestCard _50_50.jpg. Use the ImageJ image calculator to subtract one from the other. What conclusion do you draw about the effect of repeatedly saving an image in the lossy jpeg format?

### 11.4.5 Activity: Lossless PNG Format

PNG (Portable Network Graphics) is one of the lossless image compression methods included in the article and it is available in ImageJ. The PNG format compression method includes Huffman coding, which was discussed in Chapter 6.

Close any open images, and open the image file EBLTestCard.tif.

Select File | Save As | PNG ..., and save the image as EBLTestCard.png. Close the image by clicking on the cross at the top right.

Select File | Open and select EBLTestCard.tif.

Select File | Open and select EBLTestCard.png.

Answer the following two questions, then close any open images.

### Question 11.4.5.1

Are there any visible or numerical differences between the two images?

### Question 11.4.5.2

Calculate the compression ratio for the PNG image using Equation 1 given in the article. Note that the original image size in this expression is the Real size before compression, and the compressed image size is the File size after compression. Compare this result with the compression ratios for lossy jpeg compression.

### 11.4.6   Activity: Compression Ratios Differ for Different Modalities

In this activity, compression of a CT image will be compared with compression of a similar MR image (Figure 11.16).

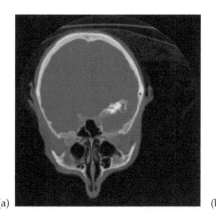 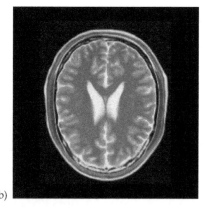

(a)                                           (b)

**FIGURE 11.16**
Images used for investigation of effect of imaging modality on compression ratio. (a) X-ray CT image (Image data from the Chapel Hill Volume Rendering Test Data Set, Volume I. [8] With permission.) (b) Magnetic resonance imaging image. (Image data from the BrainWeb Simulated Brain Database [5,6]. With permission.)

Start ImageJ.

Select File | Import | Raw …

Choose CThead.raw. Complete the Import dialog. The image is an 8-bit image, 256×256 pixels in size. The offset to the first image is 0, there is one image and the gap between images is 0. White is not zero, the byte order is not little-endian and all files in the folder should not be opened. Click on OK.

Select File | Save As | PNG …, and save the image as CThead.png. Close the image by clicking on the cross at the top right.

Import CThead.raw again.

Select Edit | Options | Input/Output.

Change the value in the box labeled "JPEG Quality (0–100)" to 100 and click OK.

Select File | Save As | Jpeg …, and save the image as CThead_100.jpg. Close the image by clicking on the cross at the top right.

Select File | Import | Raw …

Choose MRhead.raw. Complete the Import dialog. The image is an 8-bit image, 256×256 pixels in size. The offset to the first image is 0, there is one image and the gap between images is 0. White is not zero, the byte order is not little-endian and all files in the folder should not be opened. Click on OK.

Select File | Save As | PNG …, and save the image as MRhead.png. Close the image by clicking on the cross at the top right.

Answer the following four questions, then close any open images.

### Question 11.4.6.1

Calculate the compression ratio for the two PNG images using Equation 1 given in the article. Note that the original image size in this expression is the Real size before compression, and the compressed image size is the File size after compression. Suggest a reason for any difference between the two results. This question is closely related to Question 11.4.1.5.

### Question 11.4.6.2

Use the ImageJ Image Calculator to reassure yourself that the PNG images are identical to the originals.

**Question 11.4.6.3**

Calculate the compression ratio for CThead_100.jpg.

**Question 11.4.6.4**

Use the ImageJ Image Calculator to compare CThead_100.jpg with the original image.

## 11.4.7   Feedback on the Image Compression Case Study

**Question 11.4.1.1**

The compression ratio is 1.77.

**Question 11.4.1.2**

In Table 9, the real size of the CT HEAD image file is 3,492,528 bytes. With a compression ratio of 1.77 this would reduce to 3,492,528/1.77 = 1,973,180 bytes. The file is much bigger than 1 MB even after compression and so it exceeds the size limit.

**Question 11.4.1.3**

Three slices.

**Question 11.4.1.4**

Yes, it would be possible to send the NM BRAIN data by e-mail. However, if the CT compression ratio is mistakenly used instead of the correct value of 4.0 for NM, it looks like the answer is no.

**Question 11.4.1.5**

The CT images have a greater dynamic range than the nuclear medicine images, so it is less likely that there will be extensive runs of the same value. This will lead to less compression. Furthermore, there may be relatively more black background in NM images than in the CT images, again leading to long runs of the same value that compress well.

**Question 11.4.1.6**

The UFA algorithm.

**Question 11.4.3.1**

The blocking effect is seen well in the two textured squares at the bottom right in the image saved at 25% quality (Figure 11.15c). Ringing can be seen between the high-contrast low-frequency stripes in the 12% image (Figure 11.15d), and along the boundary between the gray scale and the row of squares at the bottom in all images except the 100% one. Blurring is most evident in the two low-contrast striped areas at the top right, and in the two textured squares at the bottom right. At 12% there is no texture and no stripe visible (Figure 11.15d). The change in pixel values is evident in the gray scale that runs across the middle of the image. At high compression, banding is apparent as fewer than the original 256 values are used.

### Question 11.4.3.2

The compression ratio for EBLTestCard_100.jpg is 288×324 / 23,513 = 3.97. The compression ratio for EBLTestCard_50.jpg is 288×324 / 9,437 = 9.89. The compression ratio for EBLTestCard_25.jpg is 288×324 / 8,061 = 11.56. The compression ratio for EBLTestCard_12.jpg is 288×324 / 6,430 = 14.51

### Question 11.4.4.1

The two images differ in size, the one that has been saved once is 9,437 bytes and the one that has been saved twice is 9,382 bytes. Repeated lossy jpeg saves will result in further changes to the data.

### Question 11.4.5.1

No, there are no visible or numerical differences between the two images. Image subtraction shows that the PNG images have the same pixel values as do the originals.

### Question 11.4.5.2

The compression ratio for EBLTestCard.png is 288×324 / 13,628 = 6.85. This is better than the compression for 100% jpeg quality and comes without the drawback of changed pixel values.

### Question 11.4.6.1

The compression ratio for CThead.png is 256×256 / 40,986 = 1.60. The compression ratio for MRhead.png is 256×256 / 50615 = 1.29, which is lower, suggesting that compression has been less effective for this image. Both images have similar amounts of background pixels, however the MR image has more detail in the soft tissue regions and a greater variety of pixel values. The PNG compression algorithm includes Huffman coding, which relies on using a short code for frequently occurring pixel values. If many different pixel values are present, this is less effective that in cases where a few values occur very frequently.

### Question 11.4.6.2

Subtraction of the PNG image from the corresponding original shows that the pixel values have not been changed.

**Question 11.4.6.3**

The compression ratio for CThead_100.jpg is 256×256 / 24,841 = 2.64.

**Question 11.4.6.4**

Subtraction of the jpeg image from the corresponding original gives a noisy image showing that many pixel values across the image have been changed by a small amount by the lossy jpeg compression. The higher compression ratio compared with PNG is of no value in image analysis because of the changed pixel values.

---

## 11.5   Image Registration

The article [9] for this case study is available online to subscribing institutions and is additionally supplied on the CD as an author's postprint version [10]. The images required for this case study are on the CD.

### 11.5.2   Questions on the Image Registration Article

The study described in the article [9,10] is an assessment of the performance of different registration methods on the same set of data. The study is also described in a later publication [11]. In this case study, however, the emphasis is on extracting information about the different methods of registration used in the study, and there is less concern with the evaluation aspects. There will be details in the article that are more advanced than material presented in this book, so the case study provides practice in extracting relevant detail from a wealth of information.

Referring to the Abstract and Introduction of the article, write down answers to the following questions.

**Question 11.5.2.1**

How is "Target Registration Error" defined in this article?

**Question 11.5.2.2**

What is the difference between a prospective and a retrospective registration technique?

**Question 11.5.2.3**

How were the gold standard data acquired for this study?

## Question 11.5.2.4

Explain why image preprocessing was necessary before the other groups involved could use the data to test their retrospective image registration techniques.

## Question 11.5.2.5

The information required to answer this question can be found in Section 2.7 of the article. For each of the following three registration techniques, identify one group that used the technique: Correlation, Surface Matching, Maximization of Mutual Information. The key to extracting information here is not to be distracted by the detailed explanation. Look out for words that are recognizable from the descriptions in Chapter 8.

### 11.5.3  Activity: Image Registration Using Mutual Information with ImageJ and TurboReg

Notice that none of the retrospective techniques included in the article involves the observer identification of just a few landmarks, which is the method that was used in the ImageJ activity in Chapter 8. In this activity, a mutual information method that uses the pixel values themselves rather than landmarks will be used. [12] Not only is this approach less subjective, it is appropriate because the two images to be registered are from different modalities. It is hard to identify anatomical landmarks in both images (for example, bone shows up bright in the CT image, but is not visualized in the MR image).

Start ImageJ and look in the PlugIns menu to check that there are options "TurboReg" and "Image CorrelationJ 1e." If either is not present, refer to the installation instructions that precede Chapter 1.

Two images, image02.bmp and image01.bmp, are supplied on the CD. Load the images, taking care to load image02.bmp before image01.bmp. Now follow the steps below to align the two images using a mutual information technique that does not rely on manually identified landmarks.

Select Plugins | TurboReg | TurboReg.

A dialog box (Figure 11.17) will open; select the options shown. When TurboReg was used in Chapter 8, the blue, brown and green graphics overlaid on the images were moved to select landmarks. This time leave the graphics in their default locations, and click the "Automatic" button.

A new window labeled "Output" appears. This is a stack* of two images; the first image is image02.bmp registered to image01.bmp (Figure 11.18).

---

* Read about stacks in the Image | Stacks section of the ImageJ documentation.

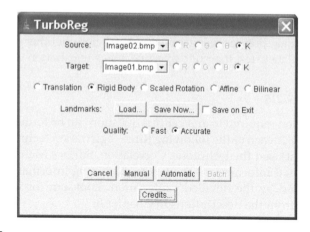

**FIGURE 11.17**
The ImageJ TurboReg dialog. (From http://bigwww.epfl.ch/thevenaz/turboreg/. [12] With permission.)

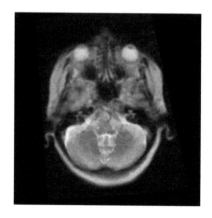

**FIGURE 11.18**
Result of automatic registration. (Image data from the Chapel Hill Volume Rendering Test Data Set, Volume I. [8] and the BrainWeb Simulated Brain Database [5,6]. With permission.)

Select Image | Stacks | Convert stack to images. This converts the stack to two separate images. Use Image | Type to convert the registered image to 8 bits, then save it.

## 11.5.4 Activity: The Joint Histogram Using ImageJ

The images are from different modalities, so performing a subtraction of the pair of registered images, as has been done in earlier activities to assess visually the success of registration, will not be informative. Instead, this is a good opportunity to demonstrate the increase in the value of the correlation coefficient for the joint histogram, which occurs when images are better registered, as discussed in Chapter 8.

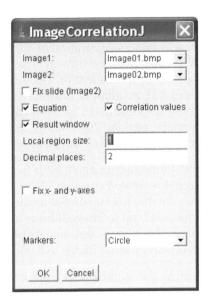

**FIGURE 11.19**
The ImageJ ImageCorrelationJ dialog.

Select Plugins | Image CorrelationJ 1e. A dialog box will appear; select the options shown in Figure 11.19 (the local region size is 1) and click on OK. Note the value for the correlation coefficient $R^2$. Close the two results windows, and repeat the CorrelationJ measurement, replacing image02.bmp with the file name of the registered result. The value of $R^2$ should increase from 0.13 to 0.48; a value of 1 is not expected in this case because the images are from different modalities.

### 11.5.5 Activity: Affine Transformation Using ImageJ and TurboReg

In the activities to date, rigid transformations have been applied. In this activity, the images to be registered do not have the same pixel size, so an affine transformation is more suitable.

Load the images image03.bmp and image01.bmp from the CD into ImageJ. Use TurboReg to register image03.bmp to image01.bmp, using an affine transformation and the accurate and automatic options. As before, convert the result stack to images, convert the registered image to 8 bits, then save it. Use Image CorrelationJ to find the correlation coefficient for image01 against image03, and for image01 against the registered result. The images are the same modality this time, and $R^2$ increases from 0.46 to 0.96.

### 11.5.6 Activity: Preliminary Alignment Using Landmarks Using ImageJ and TurboReg

To emphasize that registration was not being performed using the landmarks in TurboReg in the activities in this chapter, the landmarks were not moved before initiation of the automatic registration. In practice, however, especially if there were a large misregistration or a large difference in scale, an initial alignment will usually be performed. This is important for techniques using maximization or minimization because they may otherwise reach local, rather than global, maxima or minima (Chapter 8). In TurboReg this is achieved by placing the landmarks as if for a manual registration, but then selecting the automatic option. The landmarks are used only for the initial alignment, after which the mutual information method takes over.

Load the images image04.bmp and image01.bmp from the CD into ImageJ. You will note that image04 is rotated through 180° compared with image01, and has a different scale. On the first run do not move the landmarks and use TurboReg to register image04.bmp to image01.bmp, using an affine transformation and the accurate and automatic options. As before, convert the result stack to images, convert the registered image to 8 bits, then save it.

Repeat, but before selecting the automatic key, adjust the landmarks so that they indicate corresponding anatomical features on the two images. The landmarks do not need to be placed very carefully, but do take care to get them placed so that they correctly distinguish each side as well as the top and bottom. Compare the resulting images. In the first case, the registration has not worked well because of the large rotational mismatch between the images. In the second case, the registration visually looks much better. Measure the correlation coefficients as before to confirm this improvement. You should find that, when landmarks were used to provide an initial registration, the correlation coefficient is comparable to the previous result using image03.

### 11.5.7   Further Questions on the Article

*Question 11.5.7.1*

What is the major difference between the image registration performed in the article and the activities performed using ImageJ and TurboReg?

*Question 11.5.7.2*

It is sometimes stated that the time taken to perform a registration, including user interaction, should be about a minute if it is to be practical for clinical use. Comment on the clinical practicality of the methods tested in the article.

### 11.5.8   Feedback on the Image Registration Case Study

*Question 11.5.2.1*

The target registration error (TRE) is defined as the difference between the position of a point in the original image and its position after it has undergone a series of transformations. The point is the centroid of a region defined in the base MR image. The first transformation uses the gold standard information to move the point into its position in the source image. The transformation found using the registration method under study is then applied, and the resulting position is used in finding the TRE.

### Question 11.5.2.2

A prospective registration technique involves the placement of fiducial markers in all the images acquired, and these markers are used as landmarks for subsequent registration. In a retrospective registration technique, no markers are placed, and registration is performed using information in the images.

### Question 11.5.2.3

The gold standard data for this study were acquired using images containing fiducial markers (implanted markers and a stereotactic frame).

### Question 11.5.2.4

Before the other groups involved could use the data to test their retrospective image registration techniques, it was necessary to remove the fiducial markers from the images.

### Question 11.5.2.5

Correlation (Van den Elsen et al., Noz et al., Woods); Surface Matching (Harkness, Hemler et al., Pelizzari, Robb and Hanson); Maximization of Mutual Information (Collignon et al., Hill and Studholme).

### Question 11.5.7.1

The main difference is the dimensionality of the registration. The activities involved 2-D slices, while the registration performed in the article was of 3-D image volumes.

### Question 11.5.7.2

The methods presented in the article mostly took considerably longer than a minute to perform, but this does not mean that they will never be of practical use. Firstly, they were still being developed and would not, at the time of the study, be at their most efficient in both manual and computational aspects. Secondly, computer speeds increase as time goes by, bringing previously unfeasible methods into practical use.

## 11.6   Visualization

The article [13] for this case study is also available online to subscribing institutions [14]. In addition, an author's postprint version is supplied on the CD. The images required for this case study are on the CD.

### 11.6.1   Questions on the Visualization Article

Read the article [13,14] and answer the following questions.

*Question 11.6.1.1*

Give two disadvantages of using radiographs to assess changes in facial shape.

*Question 11.6.1.2*

List three features of the surface laser scanning technique that are suggested in the article, which make it attractive as a method for demonstrating changes in facial shape.

*Question 11.6.1.3*

Does the article give numerical values, or citations to sources containing numerical values, for the accuracy and precision of surface laser scanning?

*Question 11.6.1.4*

Is the matching procedure that was used best described as surface matching, chamfer matching or landmark matching?

*Question 11.6.1.5*

Is the transformation applied to achieve registration a rigid, affine, projective or curved one?

*Question 11.6.1.6*

When pre- and postoperative scans have been registered to each other, color-coded images are calculated to show the changes caused by surgery. Specific cases are shown in Figure 3 of the article (Figures 5 and 6 in the postprint version). Explain the radial color scheme, which is one of those used in this article. The explanation should include information on how the distance between surfaces is defined and the meanings of the colors used.

### Question 11.6.1.7

Would the colored renderings shown in the article be described as scene-based or object-based renderings?

### Question 11.6.1.8

What method has been used in these renderings to suggest depth?

## 11.6.2 Activity: Volume Rendering Using ImageJ and VolumeJ

Start up ImageJ and look in the PlugIns menu to check that there is an option "VolumeJ" there or in the list that appears when you highlight "BIJ_plugins." If it is not present, refer to the installation instructions that precede Chapter 1.

Load the data file CTheadsagy.tif,* which is on the CD. When the file is loaded, scroll through the 256 slices in the stack to see what the slices look like.

Select File | Open | CTheadsagy.tif.

Select Plugins | BIJ_plugins | VolumeJ.

Set the controls to the values shown in Figure 11.20, then click on the "Render" button. Make sure the Enter/Return button is used after changing the value in the box, or the change may not register. Rendering may take quite a long time. Bring the ImageJ command window to the front to see progress displayed as a progress bar and percentage.

The result should appear as in Figure 11.21. If it doesn't, check that all the parameters were set as shown, in particular "Rotation" and "Classifier threshold." Sometimes, after a few runs, the Render button might not start the rendering calculation (no progress bar appears). If this happens, close and restart ImageJ.

The light source in this case was set at location {0, 0, −1000}. The rendering is shown with shading that corresponds with the object being illuminated from a long way behind the viewer's head. Areas of the surface that are angled away from the viewer are a darker shade. Similarly, areas that would not receive much light from that direction look dark. The light may be moved to other locations to change the appearance of the rendering. For example, Figure 11.22a shows the result of using {0, 1000, −1000}, which is above the viewer's location, for the light settings.

---

* Image data from the Chapel Hill Volume Rendering Test Data Set, Volume I. [8] With permission.

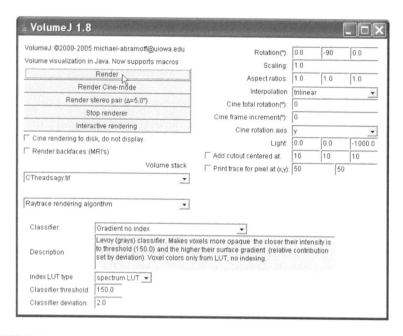

**FIGURE 11.20**
The ImageJ VolumeJ dialog. (From http://bij.isi.uu.nl/. [15] With permission.)

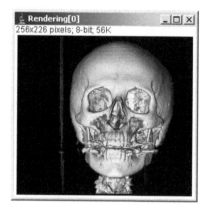

**FIGURE 11.21**
Result of rendering using the settings shown in Figure 11.20. (Image data from the Chapel Hill Volume Rendering Test Data Set, Volume I. [8] With permission.)

Keep all the other settings the same, and generate a rendering with the "Light" settings at {−1000, 0, 0}, which is from the viewer's left (Figure 11.22b).

The results so far were obtained using a "Classifier threshold" of 150.00. Before reading the next paragraph, decide if the renderings using this threshold show the skin surface or the bone surface. Should the classifier threshold be higher or lower to render the other surface of the two?

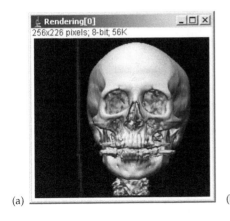

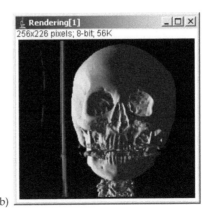

(a)                              (b)

**FIGURE 11.22**
Effect of the Light settings on rendering. (a) {0, 1000, −1000} (b) {−1000, 0, 0} (Image data from the Chapel Hill Volume Rendering Test Data Set, Volume I. [8] With permission.)

The classifier threshold of 150.00 worked well to render the surface of the bone, which appears brighter in these CT images than skin and soft tissue. A lower classifier threshold should be selected to show the skin surface.

Return the "Light" setting to {0, 0, −1000}. Keep the other settings the same as before, and generate two renderings with the "Classifier threshold" settings at 90, and then at 200. Hit the Enter/Return button when changing the value in the box. Figure 11.23a–c shows how the choice of threshold determines which surface is rendered. A lower threshold shows the skin surface, and a higher one results in thinner areas of bone being excluded. The choice of threshold is the most important one that is made in 3-D rendering. Sometimes small differences can have a large effect on the appearance of the rendering, especially if the source data have only a small dynamic range.

### 11.6.3    Activity: Generating an Anaglyph Using ImageJ and VolumeJ

Return the classifier threshold to 150.

Choose the final option in the Classifier drop-down menu: Gradient w/ depth cueing and index(spectrum).

Instead of clicking on the Render button, click on Render stereo pair button.

This will generate two images, viewed from angles separated by 5°. One will be labeled "Left eye" and one "Right eye." Bring the main ImageJ command window to the front.

Click to select the "Left eye" image, then select Image | Type | 8-bit.

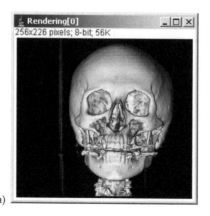

(a)

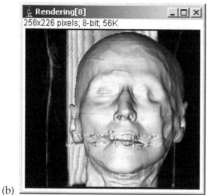

(b)

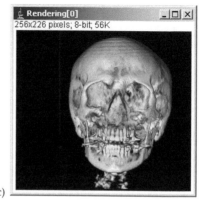

(c)

**FIGURE 11.23**
Effect of the Classifier threshold setting on rendering. (a) 150 (b) 90 (c) 200. (Image data from the Chapel Hill Volume Rendering Test Data Set, Volume I. [8] With permission.)

Click to select the "Right eye" image, then select Image | Type | 8-bit.

Select Image | Color | RGB Merge ...

Select from the drop-down menus in the dialog box, so that the Left eye image is in the box labeled "Red," the Right eye image in the box labeled "Green" and *None* is selected in the box labeled "Blue." Add a tick next to "Keep source images," and click on OK.

The result, shown in Figure 11.24, is an image with a yellowish tinge, which has a real 3-D effect when viewed with red-green glasses. Look inside the eye sockets.

Ensure that the JPEG quality setting in ImageJ is 100% (Edit | Options | Input/Output), especially if the compression case study has recently been done.

Click to select the red/green rendered image and select File | Save As | Jpeg .... Choose a directory and file name, and save the image.

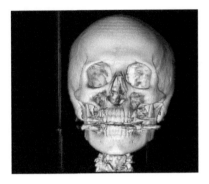

**FIGURE 11.24 (Please see color insert following page 16)**
Result of rendering a stereo pair and generating a red-green anaglyph. (Image data from the
Chapel Hill Volume Rendering Test Data Set, Volume I. [8] With permission.)

### 11.6.4  Activity: Volume Rendering Movies Using ImageJ and VolumeJ

Close the rendered images that are currently open, but keep CTheadsagy.
tif open.

Set the volume render controls to those shown in Figure 11.25 and click on
the "Render Cine-mode" button. Renderings from a total of 10 different
angles will be calculated. The images will be put into a stack. It will take
several minutes for the 10 images to be calculated.

View the sequence using Image | Stacks | Start Animation.

Stop the sequence using Image | Stacks | Stop Animation.

To save the movie to view again in ImageJ use the tif format. To make a movie
file that will run in another movie-playing program, change the image type
to 8-bit and then save as an AVI file.

For an even more impressive movie, set the cine frame increment to 10°
instead of 36°. This leads to renderings from 36 different angles and could
take up to 45 minutes to calculate on a slower computer.

### 11.6.5  One Last Question on the Article

#### Question 11.6.5.1

It was demonstrated in this activity that the CT data could be used to give a
3-D rendering of the skin surface, similar to that from the laser scanner. CT
scanners are more common in hospitals than laser scanners, and show both
bone and soft tissue, so why are CT scanners generally not used for assessment of facial surgery?

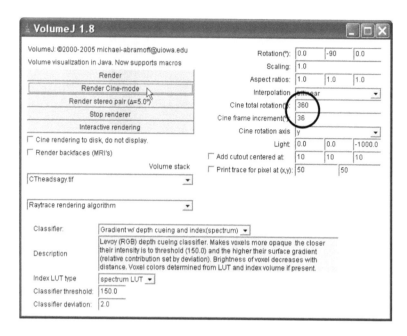

**FIGURE 11.25**
The ImageJ VolumeJ dialog, with settings changed to generate a movie sequence. (From http://bij.isi.uu.nl/. [15] With permission.)

## 11.6.6   Feedback on the Visualization Case Study

### *Question 11.6.1.1*

The two disadvantages of using radiographs to assess changes in facial shape are that radiographs have 2-D rather than 3-D information and have an associated radiation dose. Manually placed landmarks are also subjective and sparse. This will make them inaccurate in fully defining the shape of the profile, and it is assumed that the segments between landmark points are straight.

### *Question 11.6.1.2*

The surface laser scanning technique is safe for both subject and operator, is a noninvasive and noncontact method that shows the soft tissue surface, is fast and inexpensive and gives data viewable as a 3-D rendering from any direction. It has adequate accuracy and precision for the task.

### *Question 11.6.1.3*

The article says that the accuracy of imaging the face with a laser scanner is 0.5 mm and gives several references. If this aspect was of interest, it would be worth following up the references to see what measurements they report. It is possible that the word accuracy has been used loosely here and that the previous studies considered the reproducibility, or precision, of scanning.

### Question 11.6.1.4

The matching procedure would be best described as a surface matching method as the forehead and face surfaces were represented by a large number of points on the surface (forming a triangular mesh).

### Question 11.6.1.5

The transformation was a rigid one. In this article the type of transformation was explicitly mentioned. Sometimes it is necessary to deduce the transformation: if scaling is included, then the transform is affine.

### Question 11.6.1.6

The color maps indicate the distance between the two surfaces along radial lines extending from the centroid of the head in the preoperative scan. The color scale runs from purple for large positive changes, through blue and green (both positive), then through yellow and orange (both negative) to red for large negative changes. Positive changes are those where movement following surgery is in the forward direction and negative changes are where movement following surgery is in the backward direction, towards the subject. Where there has been no change the ordinary gray color of the rendering is used. The article does not define the meaning of no change, but the color scale in Figure 2 shows that it is ±1 mm. It is suggested in the discussion that this region should perhaps be ±2 mm to account for errors introduced through lack of reproducibility and errors in registration.

### Question 11.6.1.7

These are object-based renderings, because the laser scan data are a set of coordinates of points on the surface of the face, which define the object. Furthermore, because there is no volumetric information to which fuzzy values could be applied, these are surface renderings, not volume renderings.

### Question 11.6.1.8

The face has been gradient shaded so that surfaces at different angles to the viewer appear bright or dark accordingly. This can only be seen where the face is not overlaid with the color map. There is also depth shading so that surfaces that are further from the viewer appear darker than those that are closer.

### Question 11.6.5.1

The main reason that CT scanners are not used for assessment of facial surgery is the radiation dose. CT scanning is also more expensive. If the CT scan was being routinely acquired as part of normal clinical care, so that radiation dose and cost were not factors, then the disadvantage is that the CT slice thickness leads to poorer spatial resolution.

---

## 11.7   Chapter Summary

In this chapter, five separate case studies were presented. Each included questions to be answered about a specified article and a series of practical activities using ImageJ. Feedback was provided. The topics covered were image enhancement, segmentation, compression, registration and visualization.

---

## References

1. Brettle, D.S. A digital frame of reference for viewing digital images, *Br. J. Radiol.*, 74, 69, 2001.
2. http://bjr.birjournals.org/cgi/reprint/74/877/69.
3. Pruessner, J.C. et al. Volumetry of hippocampus and amygdala with high-resolution MRI and three-dimensional analysis software: Minimizing the discrepancies between laboratories, *Cerebr. Cortex*, 10, 433, 2000.
4. http://cercor.oxfordjournals.org/cgi/content/full/10/4/433.
5. The BrainWeb Simulated Brain Database. www.bic.mni.mcgill.ca/brainweb/.
6. Collins, D.L. et al. Design and construction of a realistic digital brain phantom, *IEEE Trans. Med. Imag.*, 17, 463, 1998.
7. Kivijärvi, J. et al. A comparison of lossless compression methods for medical images, *Comput. Med. Imag. Graphics*, 22, 323, 1998.
8. The Chapel Hill Volume Rendering Test Data Set, Volume I, Department of Computer Science, University of North Carolina. Available at http://education.siggraph.org/resources/cgsource/instructional-materials/volume-visualization-data-sets.
9. West, J. et al. Comparison and evaluation of retrospective intermodality image registration techniques, *Medical Imaging 1996: Image Processing, Proc. SPIE*, 2710, 332, 1996.
10. The Retrospective Image Registration Evaluation Project. http://insight-journal.org/rire/index.html.
11. West, J. et al. Comparison and evaluation of retrospective intermodality image registration techniques, *J. Comp. Assist. Tomog.*, 21, 554, 1997.
12. Thévenaz, P., Ruttimann, U.E., and Unser, M. A pyramid approach to subpixel registration based on intensity, *IEEE Trans. Image Processing*, 7, 27, 1998.

13. Miller, L., Morris, D.O., and Berry, E. Visualizing three-dimensional facial soft tissue changes following orthognathic surgery, *Eur. J. Orthod.*, 29, 24, 2007.
14. http://ejo.oxfordjournals.org/cgi/reprint/29/1/14.
15. Abràmoff, M.D. and Viergever, M.A. Computation and visualization of three dimensional motion in the orbit, *IEEE Trans. Med. Imag.*, 21, 296, 2002.

# 12

## For Instructors

## 12.1 Introduction

This chapter contains information to support the personalization by instructors of the materials presented in the book. There are suggestions for adapting and extending the activities in the earlier chapters. Some of the ideas would be best undertaken with the instructor present, and some are suitable for groups of learners to do together. The material in this chapter is intentionally more open ended than that in the rest of the book, as it is designed to be inspirational rather than prescriptive.

## 12.2 Writing Macros and Plugins for ImageJ

### 12.2.1 Macros

A macro is a script file that allows a series of ImageJ commands to be executed in sequence, and if desired, without any user interaction. Macros are very useful when an analysis protocol is under development, as they allow sequences of commands to be repeated with little effort. Once an analysis protocol is complete, a macro will ensure that the protocol is always performed in the same way. Macros are simple to write, requiring only a basic level of programming knowledge (use of loops, etc). It can be helpful to use the ImageJ record function (Plugins | Macros | Record ...), which allows the user to see how menu selections would be used in a macro. The record function may be used to save a rough version of the sequence of commands, and the saved macro can then be edited using Plugins | Edit. The ImageJ editor is self-contained: the macro can be saved and run from the editor.

Not every function that is needed in a macro can be recorded via the ImageJ menus, but there is a list of all available macro functions on the ImageJ website. [1] The list is updated frequently, and includes information about which version of ImageJ is required for the function to work. The website also includes some example macros, which are helpful for the beginner. As with

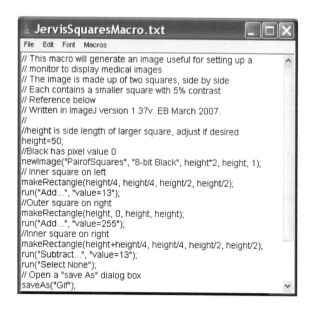

**FIGURE 12.1**

A macro shown in the ImageJ editor.

any programming, it is sensible to add copious comments, especially if the macro will be used just once a year.

One reason you may choose to write a macro is to generate test images that are free from copyright concerns; ImageJ macros were used to generate the images in Figures 11.1a and 11.14. An example of a macro displayed in the ImageJ editor is shown in Figure 12.1. In this case, the record function was used to generate an initial sequence of commands. The macro was then edited to include the variable "height" instead of the rough absolute values that had been recorded by interactively placing the regions of interest. This macro is part of an activity suggested later in this chapter.

A second motivation for using macros is that they can be used to allow students new to image processing to perform a realistic image processing protocol, without having to implement the protocol themselves. The focus of the learning may then be on the analysis of the results. Macros of this kind work well for group learning activities.

### 12.2.2 Plugins

Plugins are rather more complex than macros as they need to be compiled. Plugins allow the user to implement their own functions and use them just like any of the built-in menu commands in ImageJ. It is helpful to have a basic knowledge of Java®* to implement simple plugins, and a strong working

---

\* Java and all Java-based marks are trademarks or registered trademarks of Sun Microsystems, Inc., Santa Clara, California, USA, in the United States and other countries.

knowledge for a more complicated one. A tutorial on writing plugins can be downloaded from the ImageJ website.[2]

Several useful plugins for performing specialized tasks such as image registration and volume rendering are included on the CD that accompanies this book. Before starting to prepare a macro or plugin of one's own, it is a good idea to visit the ImageJ website to see if a relevant plugin has already been written. New contributions continue to be added to the list.

## 12.3  Article-Based Case Studies

### 12.3.1  Choosing a Case Study Article

Only a small number of article-based case studies is included in this book, and the active instructor is likely to need more, perhaps for assessment purposes or concerning a different topic area. The first point to bear in mind is that the best articles for this purpose will not normally be from the cutting edge of image processing research. Such articles are not only too focused, they are written for an expert reader who is confident in the field. A review article, a clinical study, or even an old article that may appear rather dated, will be much more suitable. Shorter articles are preferred, so a good strategy is to seek out the four-page conference paper written by the authors of a subsequent, full-length article. If a short article is not available, direct the reader to the relevant parts of the article. When searching for an article, it is usually helpful to have formed preliminary ideas about the practical image processing activities that learners will be undertaking as part of the case study. The activities will necessarily be relatively simple and suited to beginners, so it is unrealistic to seek an article that matches the activity exactly. Look instead for an article that has something in common with the practical work.

### 12.3.2  Alternative Articles for Case Studies

With experience, articles with potential to make an interesting case study will be recognized by the instructor and a list of candidate articles soon builds up. A few suggestions to start the list are given here, grouped by topic: basic principles [3], segmentation and classification [4,5], spatial domain filtering [6,7], frequency domain filtering [8], image arithmetic [9], image compression [10,11], image restoration [12,13], image registration [14,15], visualization [16,17], and good practice [18,19].

### 12.3.3  Non-medical Imaging Case Studies

This book uses medical imaging examples throughout. However, the book is also suitable for those learning image processing for another field of application. In this case, the first self-assessment question (SAQ) in each chapter requires background medical imaging knowledge, and so should be ignored, and the instructor will probably choose to replace the case studies in Chapter 11 with more relevant articles.

## 12.4 Extensions to Material in Preceding Chapters

Suggestions for extensions to the preceding material are presented in subsections that correspond with the chapters of the book. There are two categories:

- Suggestions for adaptations and extensions to ImageJ activities presented in the earlier chapters.
- More challenging ImageJ activities suitable for use when the instructor is present, or for a learner who quickly completed the earlier activities. These are presented in a style that assumes basic competence with ImageJ.

Some of the suggestions are suitable for use when working with groups.

### 12.4.1 Basic Image Processing

Look-up tables (LUTs) were introduced in Chapter 1, before the reader had developed many skills in using ImageJ. Once competent with ImageJ, it is satisfying to generate LUTs, and building LUTs is a good way to understand how they work. The raw LUT file must be 768 bytes long and contain 256 red channel, 256 blue channel and 256 green channel values. The files may be developed in a spreadsheet package such as Excel®*. The first column contains the red values, the second blue and the third green. Save the file as a .txt file in the ImageJ/luts folder. On restarting ImageJ, the new LUT will be listed with the others in the Image | Look-up Tables menu.

The Image Enhancement case study in Chapter 11 can easily be extended to include image processing operations not covered in the original. A further extension is to include different clinical examples in which overprocessing is a possibility.

### 12.4.2 Segmentation and Classification

The segmentation activities in Chapter 2 use only freehand selections and gray-level thresholding, so an extension to investigate the magic wand and the spline fitting options in ImageJ could be made. It would also be interesting to perform image classification using a clustering plugin available from the ImageJ website.

---

* Registered trademark of Microsoft Corporation, Redmond, Washington, USA.

Segmentation is an excellent topic for a group activity. The larger number of participants means that several images may be analyzed in a reasonable period of time, and a discussion of interobserver variability can be initiated. An enjoyable activity using public domain images can be based on the article by Chaudhuri et al. [20], which is about the detection of blood vessels in retinal images. The instructor will need to spend some time preparing materials, but the results of the preparation can be used, with minor alterations, more than once. A macro should be written for the learners to use, incorporating spatial domain filter kernels generated by the instructor. Reading and interpreting the article can form part of the activity, but it is usually more effective for the instructor to explain the principles behind the method before practical work begins. One approach is to divide the group in two and provide each subgroup with a slightly different version of the macro so that they can compare the effect of a difference in one part of the image processing protocol. Retinal images are available in online databases [21,22]. The databases conveniently provide gold standard segmentations, which are valuable for the evaluation of the segmentation method.

### 12.4.3 Spatial Domain Filtering

In Chapter 3 the activities concentrated on the built-in ImageJ filter kernels. Information on using the convolver dialog window to enter convolution kernels was introduced in Chapter 4. Once use of the convolver dialog has been mastered, the learners can reinforce their understanding of Chapter 3 by applying the kernels presented in Equations 3.4 through 3.7. The expected action of directional filters can be hard to grasp, and a useful extension is an exercise to clarify the relationships between the direction of gray-level gradients in the image, the orientation of the edge in the image and the action of the filters.

The unsharp mask filter is a commonly used filter that includes convolution filtering and image subtraction. It could usefully be investigated using the test image associated with the image enhancement case study of Chapter 11. Alternatively, this is an opportunity to introduce images from the instructor's specialty.

The following activity is designed as an extension to the brief mention of adaptive filtering in Chapter 3. Adaptive smoothing filters are designed to avoid smoothing across edges in an image. Firstly, the variance is measured in a part of the image known not to contain edges (Vn). The ratio (R) of Vn to the local variance at each pixel is found. If there is no edge present, then R will be close to 1. At an edge, the higher local variance leads to a small R. The principle of adaptive smoothing algorithms is to incorporate R in the calculations in a way that means the smoothing operation is only performed for pixels where R is close to 1.

## Activity: The underlying principle of adaptive filtering

In ImageJ, select File | New | Image and generate an 8-bit , 512×512 image, filled with Black.

Process | Math | Add, enter a value of 128 and click on OK.

Process | Noise | Add Specified Noise, set the standard deviation to 20 and click on OK.

Select the rectangular or elliptical selections button, and draw a shape on the image. The shape should occupy up to a quarter of the image area.

Process | Math | Add, enter a value of 20 and click on OK.

Add another shape to the image, but this time add a value of 100.

Edit | Selection | Select None.

This is now a noisy image with some edges present (Figure 12.2a). Save the image as noisyimage.tif.

Image | Duplicate.

Process | Filters | Variance and enter a radius of 2 pixels.

The result is an image with pixel values related to the variance at each pixel location (Figure 12.2b).

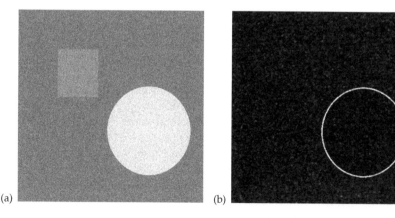

(a)                     (b)

**FIGURE 12.2**

Images for the adaptive filtering activity. (a) Example of a noisy image with edges. (b) Result of using the variance filter on the noisy image.

Ensure that the original noisy image can be seen alongside the calculated variance image. There are two things to note about the variance image. Firstly, away from the edges the image has a uniform, if noisy, appearance. The gray levels are the same over all parts of the image, including those that were brighter or darker in the original. Thus, R for these regions is close to 1. Secondly, the variance is higher at the edges of the objects in the image. This means that, as required for adaptive filtering, R will be small.

### 12.4.4  Frequency Domain Filtering

The frequency domain filtering activities in Chapter 4 included low-pass, high-pass and band-pass filtering. The band-stop filter, in which frequencies within a selected range are removed from the spectrum, was not included. Band-stop filtering requires the use of the following editing option. First draw a selection centered on the spectrum, with the frequencies to be removed outside the selection. Select Edit | Selection | Make Band ..., and enter a width in pixels. A second selection will be drawn outside the first. Use Edit | Clear to delete the spectrum between the two shapes. Remember to use Select None to remove the selections before performing the Inverse FFT.

The activities involving the removal of stripes from an image using frequency domain filtering can be extended to include examples with a range of stripe widths. These will emphasize the different locations in the spectrum of features of different frequencies in the spectrum.

### 12.4.5  Image Analysis Operations

The ImageJ Calculator was used in Chapter 5 to generate a reference image containing areas with known pixel values. The activity could be adapted to use the operations in the Process | Math menu in ImageJ, perhaps together with shapes generated using the selection tools.

The Skeletonize and Distance Map options in the ImageJ Process | Binary menu, which were not included in Chapter 5, are worth investigation. When applied to a binary image, the distance map option generates a Euclidian distance map in which each foreground pixel is replaced with a value equal to the distance of the pixel from the nearest background pixel (Figure 12.3a). An extension is to generate an image where both the foreground and background pixels are replaced with values giving the distance of the pixel from the boundary of the binary object (Figure 12.3b). If unexpected results occur, ensure that the settings for foreground and background are consistent in Edit | Options | Colors and in Process | Binary | Options .... This activity could be combined with the construction of a look-up table, which uses contrasting colors for adjacent gray values for displaying the distance map.

The red/green LUT that is supplied with ImageJ has values 0 to 127 in shades of red, and 128 to 255 in shades of green. The red/green LUT is potentially useful for displaying a registered pair of images, or perhaps to show calculated values on an anatomical background image. The desired

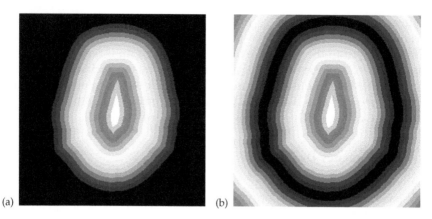

**FIGURE 12.3**
(a) Distance map of foreground of a binary image. (b) Distance map indicating distances from the edge of the binary object. This was achieved by adding the distance maps obtained from the binary image and an inverted copy of the binary image. The images were rescaled to enhance the contrast and are displayed using the 16_colors look-up table.

combination of images is achieved by masking using image arithmetic and binary images. Masking is an approach with many other applications, and it is introduced in the following activity.

### Activity: Image Masking in ImageJ

Choose a pair of different images of the same subject, perhaps a registered pair of anatomical and functional images.

Rescale the gray values of the first image (henceforth called the red image) into the range 0–127 by using the Process | Math menu to divide the image by 2.

Rescale the gray values of the second image (the green image) by dividing by 2 and adding 128.

Click on each image in turn and use Image | Look-up tables | Red/Green to apply the LUT.

Use Image | Adjust | Threshold on the red image, and adjust the sliders so that the areas to be retained are highlighted. Apply the threshold. This generates a mask with values 0 and 255. Subtract the mask from the green image (do not select the option for a 32-bit result).

Click on the mask image, and select Image | Duplicate ….

Select Edit | Invert to create a second mask with inverted gray levels compared with the first.

Subtract the second mask from the red image (do not select the option for a 32-bit result).

Add the masked red and green images.

### 12.4.6  Image Data Formats and Image Compression

The short introduction to DICOM®* format images in Chapter 6 can be extended using one or more data sets from the collections of DICOM images that are available on the Web. A search using the terms "DICOM" and "sample" will be adequate to find active resources, and there are images in the Chapter06 folder on the CD. Select a file that is of interest, and also of a size amenable to the speed of Internet access available to your students. The file will be read by ImageJ without additional plugins if it is uncompressed and has folders of files with the .dcm extension. To open all the .dcm files in a folder, select File | Import | Image Sequence …. Click on the first file name in the folder and the ImageJ Sequence Options dialog will be displayed. Click OK. Questions can be devised based on the contents of the header, for example, concerning voxel dimensions, equipment manufacturer and acquisition details. The activity might be combined with a visualization activity, in which knowledge gleaned from the files is needed to generate a correctly proportioned rendering. Correct interpretation of the slice thickness and slice location is most difficult when slices are noncontiguous or overlapping.

An extension to the image enhancement case study of Chapter 11 can be made to change the emphasis of the practical activity to image compression. The case study uses a reference image which is designed to be sensitive to filtering operations, but has not been set up to demonstrate the effects of jpeg compression. To extend the case study, learners are asked to consider the list of artifacts arising from jpeg compression (blocking, color/gray-level subsampling, ringing, blurring), and based on this information to suggest new features, sensitive to compression, which should be included in the reference image. If the group members have the necessary skills, they could be asked to add the features to the image themselves. Otherwise, a previously adapted reference image will be tested using a high degree of jpeg compression.

### 12.4.7  Image Restoration

An extension to the treatment of the MTF in Chapter 7 is to calculate values and plot an MTF from an image of an imaged test object. The image could be supplied by the instructor, or be generated by the students using an image of a point spread function and the option for frequency domain convolution in ImageJ. Activities to perform correction of geometrical distortion could be developed using a warping plugin available from the ImageJ Website.

---

* DICOM is the registered trademark of the National Electrical Manufacturers Association for its standards publications relating to digital communications of medical information.

### 12.4.8   Image Registration

The article by van Herk et al. [14] was suggested as an alternative case study article on registration because of its clinical relevance. The article includes a registration technique, chamfer matching, which was not described explicitly in Chapter 8, but falls neatly into the four-step description of image registration. The features that are matched are numerical values at each pixel representing the distance of that pixel from the boundary of the object to be registered—a distance map. The transformation calculated from aligning the distance maps (and thus the object boundary) is then applied to the full data set. In the following activity, chamfer matching is demonstrated by generating distance maps and registering them.

#### Activity: Image Registration by Chamfer Matching in ImageJ

Load the images cthead_1.tif and cthead_2.tif into ImageJ. The two images are not in spatial alignment.

Use Image | Adjust | Threshold … to create a binary version of each image, where the skin surface is the boundary of the thresholded object.

Select Process | Binary | Distance Map to convert each binary image into a distance map.

Select Plugins | TurboReg | TurboReg.

Use the rigid body option, select the box to save landmarks on exit and select accurate quality. Register the pair of distance maps using the automatic option. Be sure to note which image was the Source (match) and which the Target (base). Save the .txt file of landmark data. This contains the information that will allow you later to apply the transformation to the original pair of images.

Select the Output image stack and use Image | Stacks | Convert stack to images to extract the registered distance map and save it.

Close any open images, then open cthead_1.tif and cthead_2.tif again.

Select Plugins | TurboReg | TurboReg.

Ensure that the Source and Target image numbers are the same as when the distance maps were registered. Use the rigid body option, but on this occasion use the Load button to load the landmark file that was previously saved. Do not select the box to save landmarks on exit. Select accurate quality. Register the images using the manual option (not the automatic option); these settings mean that the transformation that was previously found will now be

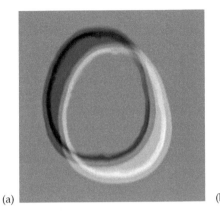
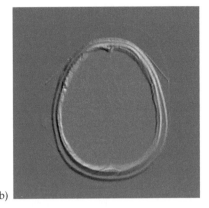

(a)  (b)

**FIGURE 12.4**
(a) Result of subtracting the unregistered pair of images. (b) Result of subtracting the pair of images after registration using distance maps.

applied to the original images. Convert the stack to images, use Image | Type to convert the registered image to 8-bit and save the image. Compare the registered image with the base image using image subtraction (Figure 12.4).

### 12.4.9 Visualization

Parametric images are an important feature in medical imaging, and images showing calculated values are used with most modalities. The instructor can generate a valuable activity by using simulation data or modified clinical data. The emphasis may be placed on either the image calculations or on production of the visualization image.

Every student will generate a red-green anaglyph for the same viewing direction if the instructions in the visualization case study in Chapter 11 are followed. A little variety may be added by asking each student to generate the view from a different angle. A reasonable spread of angles can be achieved by suggesting that each person multiply the last digit of their institutional identification number by 10 to obtain their personal angle. The visualizations generated in this way may be uploaded to a virtual learning environment (VLE), and a gallery created to display them. The case study activity may be extended to use a different visualization plugin, such as the one that was used to generate Figure 9.16.

### 12.4.10 Good Practice and Evaluation

Good practice is a topic well suited for tackling in a group learning situation. Instructors can begin the session by giving examples from their own experience, which helps to bring points to life. Guidelines for good practice in the related field of cell biology are available online [18], and these guidelines provide a basis for discussion. Learners can identify points most relevant to their own field of image processing, and can point out important issues for medical image processing that are not included in the guidelines. It is also valuable to compare and contrast the editorial policy regarding digital images in the Instructions to Authors in the *Journal of Cell Biology* [23] with the policy for the instructor's favorite medical imaging journal. As a follow-on task, learners might be asked to prepare their own adapted guidelines for good practice, suitable for use in their own professional area.

A segmentation activity suitable for groups was outlined earlier. If time permits, a comparison of the results with the relevant gold standard segmentations, leading to the construction of an ROC curve, is instructive. The group can also discuss if it would be appropriate to use filters with the same values for parameters as those in the article. A comparison of a number of different retinal blood vessels segmentation algorithms was performed by Niemeijer et al. [6]. A related activity is to obtain two of the source articles cited in that paper and compare and contrast their approaches to the evaluation of the segmentation.

An article by Jervis and Brettle [24] describes a simple way to set up ordinary computer monitors so that they all perform in a comparable fashion, using a small part of a larger test object. A pared-down test pattern can be generated in a few steps using ImageJ (a macro for generating the pattern appears in Figure 12.1). This activity will allow learners to practice using image arithmetic, while the principles behind setting up monitors using test images are suited for group discussion.

## 12.5 Publicly Available Data

If learners are to be provided with image data for practical image processing work, it is preferable to use medical image data that are already in the public domain, to avoid any doubt about privacy and ethical issues. Public domain resources include retinal image databases [21,22], a mammographic image database, [25] a variety of data for volume visualization, [26] and large DICOM datasets. [27] The image databases have usually been developed as a research resource, but in many cases their secondary value for teaching is recognized. It may be necessary to register on the website. Simulated images are valuable, especially when there is the possibility to adjust the simulated acquisition as desired, as in the BrainWeb MRI Simulated Brain Database [28].

## 12.6 Chapter Summary

Topics relevant to an instructor leading a course based on this book were covered in this chapter. ImageJ macros and plugins were discussed, and suggestions made regarding alternative case study articles. Extensions to the activities in each chapter were outlined. A short list of sources of publicly available image data was provided.

---

# References

1. List of ImageJ macros. http://rsb.info.nih.gov/ij/developer/macro/functions.html.
2. Writing ImageJ Plugins - A Tutorial. http://rsb.info.nih.gov/ij/developer/index.html.
3. Kamm, K.F. Technical note: Contrast, *MedicaMundi*, 42, 41, 1998.
4. Vannier, M.W. et al. Multispectral analysis of magnetic resonance images, *Radiology*, 154, 221, 1985.
5. Niemeijer, M. et al. Comparative study of retinal vessel segmentation methods on a new publicly available database, *Medical Imaging 2004: Image Processing, Proc. SPIE*, 5370, 648, 2004.
6. Behrenbruch, C.P. et al. Image filtering techniques for medical image post-processing: an overview, *Br. J. Radiol.*, 77, S126, 2004.
7. Muller, R.D. et al. Frequency-filtered image post-processing of digital luminescence radiographs in pulmonary nodule imaging, *Clinical Radiology* 51, 577, 1996.
8. Helenon, O. et al. Artifacts on lung CT scans: Removal with Fourier filtration, *Radiology*, 171, 572, 1989.
9. Creamer, P. et al. Quantitative magnetic resonance imaging of the knee: a method of measuring response to intra-articular treatments, *Ann. Rheum. Dis.*, 56, 378, 1997.
10. Good, W.F. et al. Detection of masses and clustered microcalcifications on data compressed mammograms: An observer performance study, *AJR*, 175, 1573, 2000.
11. Newsom, R.S.B. Effect of digital image compression on screening for diabetic retinopathy, *Br. J. Ophthalmol.*, 85, 799, 2001.
12. Velthuizen, R.P., Clarke, L.P., and Lin, H. Evaluation of non-uniformity corrections for tumor response measurements, *Engineering in Medicine and Biology Society, 1997, Proceedings of the 19th Annual International Conference of the IEEE*, 2, 761, 1997.
13. Dougherty, G. and Kawaf, Z. The point spread function revisited: image restoration using 2-D deconvolution, *Radiography*, 7, 255, 2001.
14. van Herk, M. et al. Automatic registration of pelvic computed tomography and magnetic resonance scans including a full circle method for quantitative accuracy evaluation, *Med. Phys.*, 25, 2054,1998.
15. Nogler, M. et al. Knee pain caused by a fiducial marker in the medial femoral condyle: A clinical and anatomic study of 20 cases, *Acta Orthop. Scand.*, 72 , 477, 2001.
16. McCance, A.M. et al. Three-Dimensional analysis techniques — Part 2: Laser scanning: a quantitative three-dimensional soft-tissue analysis using a color-coding system. *Cleft Palate Craniofacial Journal*, 34, 46, 1997.

17. Gehrmann, S. et al. A novel interactive anatomic atlas of the hand, *Clin. Anat.*, 19, 258, 2006.
18. Cromey, D.W. Digital Imaging: Ethics. Available from http://swehsc.pharmacy. arizona.edu/exppath/micro/digimage_ethics.html.
19. Wang, Y.-X. J. Medical imaging in pharmaceutical clinical trials: What radiologists should know, *Clin. Radiol.*, 60, 1051, 2005.
20. Chaudhuri, S., et al. Detection of blood vessels in retinal images using two-dimensional matched filters, *IEEE. Trans. Med. Imag.*, 8, 263, 1989.
21. The DRIVE retinal image database. www.isi.uu.nl/Research/Databases/ DRIVE/ (Utrecht University).
22. STructured Analysis of the Retina (STARE). www.ces.clemson.edu/~ahoover/ stare/ (Clemson University).
23. *Journal of Cell Biology* JCB Instructions to authors. www.jcb.org/misc/ifora. shtml.
24. Jervis, S.E. and Brettle, D.S. A practical approach to soft-copy display consistency for PC-based review workstations, *Br. J. Radiol.*, 76, 648, 2003.
25. The Digital Database for Screening Mammography (DDSM) http://marathon. csee.usf.edu/Mammography/Database.html (University of South Florida).
26. Volume Visualization Data Sets. http://education.siggraph.org/resources/ cgsource/instructional-materials/volume-visualization-data-sets (Association for Computing Machinery's Special Interest Group on Graphics and Interactive Techniques, originally from University of North Carolina, Chapel Hill).
27. DICOM Files. http://pubimage.hcuge.ch:8080/.
28. BrainWeb: Simulated Brain Database. www.bic.mni.mcgill.ca/brainweb/ (McGill University).

# Index

## A

Accuracy, 59, 217; *see also* Diagnostic accuracy
Active Contour Model (ACM), 48
Active Shape Model (ASM), 48–49
Actual dynamic range, 19
Adaptive; *see also* Global; Local
  compression, 134
  filtering, 74, 271–273
Addition, 102
  ImageJ activity, 110
  image plus constant, 103
  image plus image, 110
Affine transformation, 167–168, 173, 176, 184
  ImageJ activity, 253–254
Aliasing, 5–7
Anaglyph, 196
  and color vision deficiency, 212–213
  generation, 259–261, 277
Analog image, 2, 3
AND
  binary images, 113, 220
  grayscale images, 118
  ImageJ activity, 115
Angiography, 194–195
Anonymization, 202, 210
Arithmetic, *see* Image arithmetic
Artifacts
  display, 136
  image processing, 96, 226, 228
  jpeg, 134–135, 229, 243–244, 275
Aspect ratio, 30, 127, 210
Audit trail, 208, 210; *see also* Provenance, image
Automatic segmentation, 41, 48–49
Available dynamic range, 19
Average filter, 69

## B

Band-pass filter, 80
  ImageJ activity, 89
Band-stop filter, 273

Barrel distortion, 33, 148, 150
Base set, 163; *see also* Match set
  choice of, 171
  in evaluation of image registration, 215, 255
  ImageJ activity, 175, 276–277
  transformation, 166, 169–171
Bias correction, *see* Gray level inhomogeneity
Binary image, 8–9, 35–36, 58; *see also* Logical operations; Mathematical morphology
  ImageJ activity, 232, 234, 276
  operations, 273
  as reference, 234–236
Bit depth, 3–5; *see also* Data type, image; Resolution
Blocking artifact, 134, 248, 275
Blurring, *see* Filter; JPEG (Joint Photographic Experts Group); Point spread function (PSF)
Brightness, 2, 3–4, 15, 32; *see also* Gray value

## C

Calibration, image
  gray level, 112
  in ImageJ, 112
  spatial, 7, 111–112
Case study, 225, 230, 241, 250, 256
CD, xv–xvi
Chamfer matching, 256
  ImageJ activity, 276–277
Checkerboard display, 178–179
Classification (labeling), 49–50, 53–54
  multispectral or cluster, 50–52
  supervised, 52, 54–57
  unsupervised, 52
Close operator, 116
Cluster analysis, *see* Classification (labeling)
CNR, *see* Contrast to Noise Ratio
Coding redundancy, *see* Redundancy

T - #0660 - 071024 - C77 - 234/156/14 - PB - 9780367452841 - Gloss Lamination